DESIGN AND BUILD:
USES AND ABUSES

DESIGN AND BUILD
USES AND ABUSES

BY

JEREMY HACKETT, FRICS, ACIArb

|L|L|P|

LONDON   HONG KONG
1998

LLP Reference Publishing
69–77 Paul Street
London EC2A 4LQ
Great Britain

EAST ASIA
LLP Asia
Sixth Floor, Hollywood Centre
233 Hollywood Road
Hong Kong

© Jeremy Hackett, 1998

British Library Cataloguing in Publication Data

A catalogue record
for this book is available
from the British Library

ISBN 1–85978–189–6

Are you satisfied with our customer service?

These telephone numbers are your service hot lines for questions and queries:

Delivery:	+44 (0) 1206 772866
Payment/invoices/renewals:	+44 (0) 1206 772114
LLP Products & Services:	+44 (0) 1206 772113

e-mail: Publications@LLPLimited.com or fax us on +44 (0) 1206 772771

*We welcome your views and comments in order to ease any problems
and answer any queries you may have.*

LLP Limited, Colchester CO3 3LP, U.K.

Text set in 10/12pt Plantin by
Interactive Sciences Ltd,
Gloucester
Printed in Great Britain by
WBC Ltd,
Bridgend, Mid-Glamorgan

ACKNOWLEDGEMENTS

This book would have been impossible without a combination of factors which provided the impetus and opportunity to put my experience of, and some of my frustrations with, "Design and Build" on paper—hopefully for the benefit of others, be they students, practitioners, contractors or building owners about to set out on a voyage of discovery in the construction process. It may also be of interest to lawyers undertaking expensive post-mortems when things have gone wrong on Design and Build projects.

Firstly, my thanks in equal measure to my publishers LLP and to my employers Grimley, International Property Advisors, who have each in their different ways encouraged me and made it possible to bring this book to fruition.

Secondly, my thanks to several professional friends, surveyors, solicitors and others who have heard my tales of woe on Design and Build and said "why don't you pull all this together and write a book on the subject?".

Some of my tales of woe pre-date my employment with Grimley, and in any event the opinions expressed herein are mine as a Chartered Surveyor, experienced over 25 years in Design and Build or variants thereof, both in the UK and overseas. The views expressed are therefore personal and cannot necessarily be construed as Grimley policy.

In the first part of this book I have focused on key issues which distinguish Design and Build from the more traditional form of contract procurement, i.e. architect-led "lump sum" contracting, where the contractor builds to the architect's design.

To make the necessary distinctions I have taken the most commonly used standard form of Design and Build contract—the JCT Standard Form of Building Contract With Contractor's Design 1981 Edition—as a point of reference, but without seeking to comment on this particular standard form on a clause by clause basis.

In all areas I have tried to put flesh on the subject by references to real-life problems—some amusing and some apparently insoluble at the time.

In the second part of this book I have attempted to draw comparisons in key areas between six other standard forms of Design and Build contract—and accordingly I am indebted to the various organisations owning the copyright for giving me permission to quote the necessary extracts from the following standard forms of design and build contract.

I therefore gratefully make the following specific acknowledgements:

1. In respect of the Standard Form of Building Contract With Contractor's Design 1981 Edition, the JCT through RIBA Publications wish it to be stated that extracts are reproduced by permission of the copyright holder—copyright RIBA Publications 1981.

2. In respect of the ACA Form of Building Agreement 1984 (Revised 1990) BPF Edition, the BPF have kindly given written consent to the extracts as reproduced.

3. In respect of the GC Works/1 (Edition 3) July 1993 Form of Contract, Her Majesty's Stationery Office have kindly given written consent to the extracts as reproduced.

4. In respect of the IChemE Model Form of Conditions of Contract for Process Plants, The Institute of Chemical Engineers as authors and publishers have given written consent to the extracts as reproduced.

5. In respect of the Engineering and Construction Contract November 1995, Thomas Telford Publishing on behalf of the author and publishers, The Institution of Civil Engineers, London 1995, have given written consent to the extracts as reproduced.

6. In respect of the ICE Design and Construct Conditions of Contract, Thomas Telford Publishing on behalf of the joint authors, The Institution of Civil Engineers, The Federation of Civil Engineering Contractors and The Association of Consulting Engineers have given written consent to the extracts as reproduced and, as publishers, The Institution of Civil Engineers, London 1992, do likewise.

7. In respect of the FIDIC Conditions of Contract for Design and Build and Turnkey 1995, the FIDIC Secretariat wish it to be stated that their Conditions are subject to copyright and may not be reproduced without written permission of the Federation, and that copies of their Conditions and "The Orange Book Guide—Guide to the use of FIDIC Conditions of Contract for Design and Build and Turnkey 1995" may be obtained from the FIDIC Secretariat, PO Box 86, 1000 Lausanne 12, Switzerland or through its Member Associations.

To all the above my grateful thanks and I trust my comments, where applicable to the various standard forms, will be read as constructive criticisms intended to lead to further understanding of Design and Build. When properly handled, Design and Build can be very, very good—when badly handled or deliberately abused it is a disaster waiting to happen.

Finally, these acknowledgements would be incomplete without a special thank-you to all at LLP, to Julian Critchlow of S.J. Berwin, Solicitors, who kindly reviewed my earlier draft—and suitably encouraged me—and to the charismatic Mark Cato. Without Mark's organisation of the annual Arbitration Club Dinner, I would not have had the good fortune to have sat next to a young lady from LLP and the seeds of this book would probably never have found fertile ground. Of such happy chances is life made.

April 1998   THE AUTHOR

NOTE:

As a matter of house style, the publishers have only given initial capital letters to proper nouns and a limited number of other terms which by convention have capital letters.

Only when quoting directly from standard contracts have the publishers given initial capital letters to defined contractual terms, e.g. Employer's Agent, Employer's Requirements, Conractor's Proposals, etc., and the author trusts that this policy decision will not detract from the understanding or clarity of the textual meaning.

TABLE OF CONTENTS

CHAPTER 1

INTRODUCTION

1.1 USAGE

It has been my professional good fortune to have worked on a wide variety of construction projects both in the UK and overseas, and to have met some fascinating characters—as well as to have been confronted with all sorts of problems arising from different procurement options.

When I have been called in many of these problems have been fairly advanced so I have been privileged to meet some amazing legal minds and to have been involved in various litigations and arbitrations as expert witness. However, in recent years design and build projects seem to have attracted more than their fair share of problems, with arbitration being the normal dispute resolution route—assuming *both* parties can afford it.

Very soon it will fall to adjudicators to dispense instant justice on a wide range of potential disputes, including design issues, with arbitration or litigation as a subsequent remedy open to an aggrieved party.

As such, there needs to be greater awareness of just how different the design and build procurement route can be in real terms as compared with the more traditional architect-led lump sum contract where the contractor simply builds out as instructed—notably in respect of the contractor's obligations and the various professional consultant's varying roles.

As a historical aside my treasured copy of *Spon's—Architects and Builders—Pocket Price Book 1934* sets out the following fee scales in relation to disputes procedures, with fees apparently being at fixed published rates:

Arbitrations, Fees payable in.

For each arbitrator or umpire:

	£	s	d
For the first hour of sitting	2	2	0
For each subsequent hour	1	1	0

	£	s	d
If case jointly stated under Rule 43, not to exceed...	4	4	0

The above scale shall apply unless the parties enter into an agreement, to be endorsed on the submission, to pay specified fees of a larger amount

Counsel and Solicitors	Fees to be on the same scale as those allowed in the High Court; but for attendances at the hearing only

It can safely be assumed that in 1934 construction industry arbitrations were very much the poor relation of litigation, being the province of non-lawyers. Just imagine counsel and solicitors not being entitled to fees for case preparation!

In 1934 a London bricklayer was paid 1/7d per hour (8 new pence) and a London labourer was paid 1/2¼d per hour (6 new pence). Equivalent London bricklayer and labourer rates in 1997 are £5.57 and £4.62 per hour respectively, so £1.1.0 per hour in 1934 would equate to only £77 per hour for the arbitrator in 1997.

Moreover, the construction world does seem to have a habit of going full circle—note the provision for an "Umpire", as an alternative to an arbitrator, and the provision for a case jointly stated, subject to a fee cap. It would appear that this pre-war practice was probably the forerunner to adjudication, which we think we invented in 1996, post the Latham Report, but in reality adjudication has its roots in commercial arbitration procedures as they evolved from the middle ages and has been written into some forms of standard building industry contracts over the last 20 years, mainly in respect of payment set-off disputes, ie the first "quick fix" procedure, subject to later appeal.

However, to revert to the main purpose of this book, the title of this book *Design and Build: Uses and Abuses* was carefully chosen to represent the three primary objectives inherent in the text:

- To give a wider understanding of design and build as a valuable procurement option.
- To give a realistic and hands-on insight into the problems and pitfalls awaiting those who stumble into design and build for the wrong reasons.

- To show how contract drafting must address the different allocations of risk inherent in design and build and how the employer's agent should be empowered to administer the contract.

The focus of this book is the most commonly used standard form of design and build contract currently used in the building industry, namely the Standard Form of Building Contract with Contractor's Design 1981 Edition, as published by the Joint Contracts Tribunal—but it is used essentially as a vehicle for identifying the fundamental principles and problem areas which can arise, rather than as a commentary on the Standard Form of JCT With Contractor's Design 1981 Contract, *per se*.

Chapters 1 to 9 therefore take the JCT With Contractor's Design 1981 Contract as a basic reference point, but in Chapter 10 the bones have been picked out of several other standard forms of design and build contract commonly used in the building, civil engineering and process engineering industries—and, interestingly, they offer some different solutions to the regular problems in various areas.

As such, this book deliberately does not follow the JCT with Contractor's Design 1981 contract clause-by-clause, but seeks to highlight the regular problem areas which are unique to the design and build concept—offering some real life tales of woe and some suggested solutions.

Going back to basics, therefore, the construction process has many variants as to what happens in the time between a landowner conceiving plans for building, whether a new building or refurbishment, and the big day when he takes occupation. Inevitably the employer has three primary concerns, in the following suggested order of priority:

1. Cost
2. Time
3. Quality

In some circumstances, such as a manufacturing process, or in the retail industry, *time* can become the top priority when the commercial cost penalty of late completion of the construction process outweighs what might be involved in paying accelerated construction costs—a good example being McDonald's, who have been known to build out restaurants in under a week but at a premium price. However, on most projects it will be *cost* that matters most, with *quality* coming a poor third behind *time*.

All building procurement systems are therefore variants of how these three primary requirements are managed and who takes the responsibility for so doing. The other side of the coin is the risk of non-management and the contractual remedies which may be extracted by the supposedly innocent party. Rarely when things go wrong will it be clear-cut as to who did not perform in the first place, so often there is not a totally innocent party and a dispute arises, with allegations and counter-allegations of default. "Responsibility" and "risks" are therefore recurring themes of this book.

Traditionally the "Architect" or "Engineer" was God in full control of the three basic stages of any construction project, ie the design, the procurement and the supervision of the works on behalf of the building owner. As anyone who has read the two classics of the construction industry, *The Honeywood File* (1929) and *The Honeywood Settlement* (1930) by HB Cresswell, will know, there was a strict social order in any building contract—the employer, Sir Leslie Brash and his dreadful wife Lady Maude, secondly James Spinlove, architect, a friend of a friend in Sir Leslie's London club, and—very much third—the contractor, Griglay. The first wheel falls off when the owner's daughter's horse breaks a leg in an unguarded trial pit and has to be put down, and then Spinlove pre-orders the bricks but a quality dispute arises when they are delivered to site. Matters go from bad to worse and solicitors are brought in at an early stage: *plus ça change*! A misunderstanding of client's brief by Sir Leslie and late design detailing then alienates Griglay and we have the classic recipe for additional *cost* and *time* claims.

The industry response, some 50 years or so since Mr Cresswell's two classic works of fiction, was to try to eliminate the potential for conflict between employer and contractor, which often arises due to questionable performance of the consultants employed by the building owner, by moving the risk of design from the employer to the contractor, assuming basic scope of work and performance parameters had first been established by the employer.

Although not a new concept in the construction industry this procurement system became known as *design and build* and the contractors who first promoted this system would either have in-house design teams, some directly employing up to 60 architects and employees with associated professional skills, or would form partnering arrangements with well-respected architectural practices.

In the 1970s and early 1980s design and build delivered value-for-money projects, mainly relatively simple, if large, industrial developments or office blocks relying on a high degree of repetition and prefabrication. Design and build as a procurement system then became more widely used on all types of projects, including complex refurbishment projects—in many cases quite inappropriately.

The recession of the early 1990s then set in and contractors generally diversified and started poaching one another's traditional work sectors. It was at this stage that abuses became commonplace and design and build began to have serious critics—just as management contracting had suffered when the genuine contractors had their reputation collectively tarnished by less principled operators entering the market in the search for easier pickings. At much the same time employers became increasingly dissatisfied with the traditional architect-led design route, with its record of overruns in both *cost and time* (which contractors translated into extension of time and loss and expense claims), perceiving design and build to offer the advantage of single-point responsibility.

In defence of architects generally, they now have an almost impossible task in that the range of services and skills they are expected to deliver on major projects can rarely be found in one individual. They are expected to deliver:

- Conceptual design
- Early cost and procurement advice
- Detailed design development
- Tender preparation and control
- Working drawings
- Ongoing cost advice
- Detailed project administration
- Close site supervision
- Full specification and statutory compliance
- Delivery of the project on cost and time.

Basic tasks, one might say, but given the complexities of modern construction and contractual arrangements the performance of the architect usually depends on team working.

It is in the area of timely delivery of drawn information that most contractors' claims originate. If one wishes to train as a naval architect one has to spend a year on the tools in a ship- or yacht-builder's yard, trying to interpret drawings with few straight lines! As such it is a fast learning curve as to what is required on a drawing for it to be intelligible and definitive. I would suggest architects in the construction industry would benefit from such practical experience at the sharp end.

So the first phase of any design and build project is all about minimising risk—more specifically, defining objectives in performance terms only and then deciding *when the design responsibilities will pass from the employer to the contractor.*

Whether specialising in design and build or not, most contractors have their shareholders to account to, and will only do their minimal best. On design and build projects this factor often manifests itself by the contractor designing down the original concept and reducing the quality of materials and workmanship—to maximise profits and provide a buffer against unforeseen site and subcontracting problems.

Once a design and build contract is awarded then there is a natural void or vacuum, i.e. there is no role for the traditional architect as retained by the employer with the dual responsibility of designer and supposedly independent certifier of *cost, time* and *quality* as under JCT 80. This void is filled under JCT 81 Design and Build by the employer's agent. He, or she, is therefore a key player and any design and build contract must be clear as to the extent of the agent's powers in acting on behalf of the employer—which are distinctly different from those of the architect or engineer under a traditional lump sum contract: see Chapter 3.

1.2 PRACTICE GUIDES AND CASE LAW

Apart from Practice Notes and Commentaries issued with the various standard forms of design and build contracts (see Chapter 10), there are only three or four existing textbooks specifically on the subject of design and build, as well as numerous articles. However, the main thrust of these points of reference for good practice appears to be "What the contract says, or should say", coupled with "This is how design and build works, or should work in practice"– but with no real life examples and anecdotes.

Theory and practice are often strange bedfellows, and despite careful contract drafting, and despite detailed interviewing of potential contractors at the bid stage etc., things can and do go wrong under design and build, just as they do under other forms of procurement.

Design and build is not a panacea for avoiding conflict and never will be if the two main parties have differing objectives and perceptions. The three essential requirements of any contract must be:

- A defined employer's brief—whether a short performance specification or fully detailed.
- Clear apportionment of contractual risks—such that they can be assimilated by the contractor and priced in his tender.
- Provisions for forward control of employer's changes—rather than retrospective argument, and with such changes being kept to manageable proportions.

The original standard JCT Standard Form of Contract With Contractor's Design 1981 replicated the thinking and change-order procedure of the better known JCT 1980 Contract, where in theory a building scheme is fully designed by the architect, with "For Construction" drawings hopefully issued to the contractor in good time prior to construction. Inevitably, either the employer or architect may wish to make some changes and variations are then issued by the architect as and when, but with no requirement for pre-costing or agreement of time implications.

Such a passive change-order procedure leaves too much to chance and depends upon the goodwill of individuals whose job it is to agree variations or changes after the event, both in *cost* and *time*.

To cover this unsatisfactory aspect in the original version of JCT 1981 With Contractor's Design, the Supplementary Provisions were brought in at the instigation of the British Property Federation, but as an optional set of clauses under Amendment 3 in February 1988. These supplementary provisions are fully discussed in Chapter 5 but essentially they were designed to give *cost* and *time* certainty, for the benefit of *both* employer and contractor when faced with the need for changes, as well as providing a first-fix dispute procedure by way of adjudication.

The Standard JCT 1981 With Contractor's Design Form of Contract then provides a formal disputes procedure by way of arbitration at Clause 39. Until

1996 and the coming of the new Arbitration Act the perceived wisdom was that arbitration was to be preferred to litigation, so rarely was Clause 39 struck out of design and build contracts.

However the standard JCT 1981 is a compromise document, negotiated by a committee of 20 or more representing all elements of the construction industry from employers to professionals to main contractors to specialist subcontractors, so inevitably it is silent or stops short of providing who should do what in certain site situations.

Because Clause 39 has generally been left in signed JCT 81 contracts there is very little case law to cover some important issues which, due to the different allocation of risks and responsibilities of JCT 1981 as opposed to JCT 1980, are particular to design and build.

Under the new Arbitration Act 1996 the late deletion from the original discussion draft of clauses 85 to 87 removes the power of the courts to refuse a stay of arbitration. In practice an aggrieved party to a final account dispute will therefore not be able to ask the court to open up and review architects' certificates etc. in contracts with an arbitration clause, even in multi-party disputes.

This is a cause for concern and some solicitors are now deleting arbitration clauses from standard contracts for this very reason. As such, more construction disputes involving design and build contracts are likely to be heard in the courts, i.e. the Official Referee's Court and consequently case law relevant to design and build problem areas is likely to evolve.

1.3 LANDMARK CASE

It is not unusual for cases to take between five and 10 years to come to court and given the recent recession it is a fact of life in the construction industry that some of the original parties to a contract, e.g. the design and build contractor, have in the meantime ceased to trade.

Given this scenario many employers are now finding themselves saddled with substantially defective buildings. Where those buildings were procured by design and build, those unhappy employers are now looking to employer's agents and project monitors for recompense, via professional indemnity insurers.

Generally speaking, professional services fee agreement do not contain arbitration clauses so I expect to see a clutch of such design and build cases coming into the Official Referee's Court in the next five years.

One such judgment has been handed down as the author was finalising the draft for this book and concerns defective roofs at the new UK manufacturing headquarters of George Fischer Holding procured by design and build in 1990 under the JCT Standard Form of Contract With Contractor's Design 1981.

It was a typically convoluted contract scenario which required His Honour Judge Hicks QC to devote several pages of his judgment to describing the "factual setting" before turning his attention to the duties and alleged breaches of the employer's agent.

By the time the case in its present form came to court in 1997 the design and build contractor had gone into liquidation and the previous in-house design consultancy had managed to buy themselves out. So the employer sought to recover from the employer's agent, Davis Langdon & Everest.

I shall therefore be referring in the course of the following chapters to this landmark case for design and build—known formally as *George Fischer Holding Limited* v. *Davis Langdon and Everest & Others* (1994) ORB 775 which, *inter alia*, touches upon the following duties of the employer's agent:

- Duty to appraise and advise on suitability of the contractor's design proposals at tender stage.
- Duty to approve and comment upon the contractor's post contract design development—an express duty in this case.
- Duty to watch the works in progress for specification and workmanship compliance, paying particular attention to high risk areas.
- Duty to bring in specialist advice where expertise required is beyond in-house skill base.
- Duty to properly certify practical completion—patent defects and the contractor not having vacated the works being key criteria in this case, irrespective of partial occupation by the employer.
- Duty to warn the employer of the commercial consequences of issuing practical completion when patent defects known to exist, with particular reference to the release of retention monies and the performance bond.

1.4 PROBLEMS IN PRACTICE

The purpose of this book is therefore to explore typical problems as the author has seen them on various design and build projects and to suggest by critical comment how they might have been avoided in the first instance, or how they might have been resolved short of arbitration or litigation. We can all learn by our own and others' misfortunes, and experience shared must be experience gained.

In mitigation I plead no formal legal training but a healthy respect—and at the same time disrespect—for the law. As a quantity surveyor who has, over 30 years, worked for professional practices, for UK and foreign government organisations, for main contractors and subcontractors, but now specialising in trouble-shooting and general litigation support work, I believe case law is a developing science. No two cases are the same in fact, in presentation or in the perception of the judge or arbitrator.

As such, existing case law is there to be challenged depending on the facts and merits of the particular arguments, and only by highlighting problem areas will practice and procedure in design and build develop to cover the gaps left by standard contracts and the abuses that inevitably follow.

The principal problem areas involved in design and build as a procurement system can be summarised as:

- *Benchmarking*—defining the quality to be delivered in the finished project.
- *Fitness for purpose*—whose risk?
- *Contract base-line*—What was allowed for in the contractor's accepted tender and how should an employer's change be valued?
- *Changes*—What is perceived by the contractor to be an employer's change may be seen by the employer or his agent to be no more than approval of the contractor's design development.
- *Delays*—Who controls the information release programme and what does the contractor do if employer approvals are delayed?
- *Certification*—who decides interim valuations, quality compliance and final account issues?

Hopefully in the course of the following chapters it will become clear by way of discussion and some real-life examples how these problem areas might be addressed, and I express my thanks to those various design and build contractors, employer's agents, and even solicitors and counsel who have in their various ways unknowingly provided the material for this book.

CHAPTER 2

THE ESSENTIAL FEATURES DISTINGUISHED

2.1 THE DESIGN TEAM AND THEIR RESPONSIBILITIES

Traditionally the design team on a building project will be headed by "the architect" and on a civil engineering project by "the engineer". Whether architect or engineer, he is responsible to the employer, who pays his fees, for the total design, supervision and delivery of the project, hopefully on time and on budget. He has a duty to keep the employer fully informed at all stages of progress and in particular if any issues arise which may prejudice the agreed budget and required timescale.

The architect or engineer will be assisted in this task by other individuals from within his own office or from outside consultancy practices and typically the design team on a medium to large building project will consist of:

- Architect
- Structural engineer
- Mechanical and electrical services engineer
- Quantity surveyor
- Clerk of works

As most contractors' claims are founded on what the architect is alleged to have done, or more likely alleged not to have done, he is in an invidious position being, as it were, asked to be judge in his own court where he is also the accused. Human nature being what it is, he has a position to defend, particularly where his alleged misdemeanours mean the employer will have to meet additional costs and/or not get his building delivered on time.

Design and build neatly avoids such conflict of interest by passing the parcel of design liability and timely production of construction drawings from the employer to the contractor. However, where the same architect or practice is then re-employed by the contractor as was originally employed by the building owner for the concept designs, design and build merely replaces one area of potential conflict of interest with another—more fully discussed later in Chapter 3 under Novation.

Thus the primary distinction of design and build is that it is the contractor, not the employer, who employs the design team post-contract, and this can quickly lead to problems and disillusionment, particularly for professionals who have previously only known life the other side of the fence!

One structural engineer recounted to me how at his first internal site meeting with his new client, a design and build contractor, he was passed the block they intended to use in order to achieve a significant cost saving over that tendered. Outwardly the block looked very similar to that tendered, but when he managed to break it over his knee he rejected it. The contractor left him in no doubt that his impromptu action and subsequent opinion was not welcome and he should confine himself to producing working drawings. He picked up his coat and left the meeting, faxing his resignation an hour later when he got back to his office.

This is a salutary tale for all professionals who for whatever reason seek employment by design and build contractors: unless you are prepared to give the contractor unbiased professional advice and stand your ground just as you would have advised the employer in the same situation, you compromise your professional standing within the team and sooner or later may well compromise your professional indemnity (PI) insurance. If the contractor prefers his own advice or simply refuses to listen then you have to be very careful to record the facts, depending upon their potential significance.

The responsibilities of the remainder of the design team are no different, in that they have a single client—the contractor. However well they know, or may have known, the employer, he is now essentially the "other side" and any unguarded comments in the pub or leaked information on the project to the employer, or his agent, would certainly be a breach of trust as between the professional and his client, the contractor.

Given that under design and build all the design team are in the contractor's camp, it poses the immediate question "Who is left to safeguard the Employer's interests?" The answer is "the employer's agent"—whose role and responsibilities are discussed in Chapter 3.

2.2   FITNESS FOR PURPOSE

I know of no other subject other than that of practical completion where I have heard such conflicting and unclear legal opinions. If I, as an end user, buy a product, say a washing machine, it matters not to me who designed it and who manufactured it. I appreciate that different companies were involved in the two distinct processes, but if it goes wrong I hold the retailer, as the manufacturer's agent, fully responsible under the Supply of Goods and Services Act 1982.

So what is different if I procure a building from a design and build contractor? Why will most design and build contractors not give me an unqualified "fitness for purpose" warranty and why does the JCT 81 Form of Contract duck the issue at Clause 2.5.1? This reads:

"Insofar as the design of the works is comprised in the Contractor's Proposals and in what the Contractor is to complete under Clause 2 and in accordance with the Employer's Requirements and the Conditions (including any further design which the Contractor is to carry out as a result of a Change in the Employer's Requirements) the Contractor shall have in respect of any defect or insufficiency in such design the like liability to the Employer, whether under statute or otherwise, as would an Architect or, as the case may be, other appropriate professional designer holding himself out as competent to take on work for such design who, acting independently, under a separate contract with the Employer, had supplied such designs for or in connection with works to be carried out and completed by a building contractor not being the supplier of the design."

I guess that this magnificent example of non-plain English (a 145-word sentence) took days, not hours, to agree in JCT committee, and to my mind it flies in the face of the whole philosophy of design and build, i.e. single-point responsibility.

Some time ago I was asked to advise a very wealthy client who had a giant-sized open-air swimming pool, but in a sloping garden, and who wanted to cover it in, given the vagaries of the English climate. His preferred contractor was advocating design and build, but when I asked the contractor "What if the existing pool goes sideways when you excavate the foundations for the wall and roof structure?" he replied it would be "extremely unfortunate" and the employer's problem. I pointed out that the employer was offering no design input and was relying entirely on his chosen design and experience as a contractor. He continued to refuse a fitness for purpose warranty. When I then enquired why he was advocating JCT 1981, as opposed to JCT 1980, he had no answer.

I would therefore submit that in respect of design and build the normal duty of care owed by an architect or other design professional is not relevant, and that in any event the definition in itself is open to a wide range of legal opinion. In another case concerning the duty of care owed by a quantity surveyor I was left in no doubt by our leading counsel, best known as a criminal lawyer rather than a construction lawyer, that the definition of professional negligence was that laid down in a medical negligence case, *Whitehouse* v. *Jordan* (1980), where a forceps delivery went wrong. In *Whitehouse* v. *Jordan* the appeal judge held that the test of professional negligence was whether " . . . he had fallen below the standard of skill expected from the ordinary competent specialist and had therefore been negligent" and decided, as did the House of Lords, on the evidence, that the surgeon might have performed better, but had committed no obvious act of negligence.

Unfortunately in the construction industry there is always some practitioner who, although generally competent, will cut corners on a particular project, so in practical application I would suggest *Whitehouse* v. *Jordan* is a very restrictive burden of proof for anyone alleging professional negligence against an architect, engineer or quantity surveyor.

Happily there are four cases more relevant to the construction industry and to design and build in particular:

- A professional designer is not required to guarantee the result unless he also delivers the end product—*Samuels* v. *Davis* (1943).
- Where the employer is clearly relying on the contractor for a contract for work and materials the law can imply a fitness for purpose obligation depending on the facts of the case—*Young & Marten Ltd* v. *McManus Childs Ltd* (1968).
- Where there is no independent designer but the contractor offers a whole building, a fitness for purpose obligation is likely to be implied and it is not necessary to prove negligence in design or fault in materials or workmanship—*Viking Grain Storage Ltd* v. *TH White Installations Ltd* (1985).
- The defence that a duty no higher than reasonable care in design was required was further rejected by the House of Lords in the case of the collapse of a television mast—*IBA* v. *EMI Electronics Ltd and BICC Construction Ltd* (1980).

Those who oppose this proposition of fitness for purpose in design and build will on the other hand no doubt be quick to quote other case law and seek to distinguish those precedents from the four leading cases quoted above.

When all is said and done, it is the employer who writes the ground rules and if the contractor does not like those ground rules he has two options: he can decline to tender or he can put a premium price on what he perceives to be an onerous contract condition.

My personal preference is to make it clear in the preliminary tender enquiry for any design and build contract that fitness for purpose will be an express term and that only contractors who can offer a performance bond and will warrant fitness for purpose will be admitted to the tender list.

The usual contractor's objection is that he cannot obtain insurance cover for "fitness for purpose" at other than a premium rate. Whilst this may have been true in the past, it is no longer true, and, in any event, the employer must be given the option to buy "belt only" or "belt and braces", so to speak.

If one then uses JCT 1981 one *must* amend Clause 2.5.1 in the signed contract or otherwise one might well be faced with a legal argument, later in the project, that Clause 2.5.1—limited duty of care—takes precedence over the "Contract Particulars" section of the employer's requirements where "fitness for purpose" has been carefully written in.

2.3   THE TENDER PROCESS

Under a traditional architect design-led contract such as JCT 80 the tender process if handled properly is time-consuming, but at least it should ensure that the tenders are being considered primarily on *cost*, with *time* usually a secondary consideration, i.e. the design is the employer's design as set out in the contract documents, and as measured in the bills of quantities, assuming quantities are applicable, and *quality* is stated in the specification and bills of quantities. Given this scenario tenders should be directly comparable with one another.

Thus there is a stated base-line for all to see, against which any variations instructed post-contract by the Architect can be considered and agreed.

Typically JCT 80 contract documents will consist of:

- Drawings—reasonably detailed
- Specification—contract conditions, materials and workmanship
- Priced bills of quantities
- JCT 80 Standard Form of Contract
- Appendices covering required forms of bond and warranties

The big difference at tender stage under design and build is of course that there is no reasonably detailed design and that there are two elements being tendered:

- Design development
- Total cost, including design fees

It follows that if there is no reasonably detailed design there can be no bills of quantities prepared by the employer's quantity surveyor. Consequently the likelihood of misinterpretation by the competing tenderers of the employer's brief is considerably greater.

This will militate against the genuine design and build contractor who speculatively invests good money at tender stage in developing the employer's outline design and then costs a fully considered scheme, as against the less scrupulous design and build contractor who will not invest much at tender stage on design development and will cost a minimalist scheme, with construction details and material selection at the lower end of the cost range of acceptable standards. Such contractors are then well practised in finding discrepancies in the contract documents and persuading employer's agents in respect of allowable changes, so recovering monies they have deliberately suppressed in order to win the project in the first instance.

It is always tempting to accept a low tender or, more cynically, not to find good reason to reject it, and this is even more relevant when considering JCT 81, as opposed to JCT 80 tenders based on the employer's bills of quantities. Moreover, in some situations, for instance involving public bodies, standard procedures require the acceptance of the lowest tender, unless there are clear grounds to disallow it—which is different from giving an opinion that it is a deliberate under-bid and therefore the "lightning before the storm".

On one particular design and build project, on which I was called in as auditor when it got into trouble half-way through the construction period, the employer's agent prepared a standard elemental cost plan totalling £5.5m, based on the known footprint area of the building. Certainly some of his elemental unit rates were nervously high but when the lowest tender came in at £3.8m the writing should have been on the wall.

Strict forward cost control and strict site supervision of *quality* as built should have been the order of the day to hold the contractor to the undoubted bargain offer—but was not. The result was that the contractor soon had the initiative and the employer had to invest very considerable fees in retrospectively arguing which items in the contractor's final account were valid Clause 12 changes and what was their proper value—apart from which there was a long list of defects at final inspection, several of them serious and costly to rectify.

So, how should a design and build tender enquiry be formulated? The basic framework is provided by the JCT 81 contract which requires in the First Recital that the employer has issued to the contractor his requirements (hereinafter referred to as "the employer's requirements"). Normally these requirements will consist of:

- Outline drawings—usually plans and elevations sufficient to obtain outline planning consent, but can be more detailed.
- Specification—setting out all required contract particulars and materials and workmanship requirements, including such matters relevant to the contractor's considerations of fire escape strategy and other building control compliance issues, e.g. number of occupants intended on each floor level.

Sometimes of course a design and build tender enquiry might be a true design competition, i.e. site known and parameters of design stated, but no preconceived building footprint or elevational detail. Such situations are rare and generally clients have reasonably developed briefs as to space planning, number of occupants, intended usage, possible longer term options, etc.

The Employer then has the problem of how to call in tenders and how to compare them on an objective basis, other than just bottom line price.

The first requirement under JCT 81 is that the contractor responds to the employer's requirements with his contractor's proposals setting out in some detail how he proposes to meet the employer's stated brief. Inevitably the contractor will use the contractor's proposals to confirm any assumptions arising from his reading of the employer's requirements and to state any qualifications, exclusions, etc.

But what if the clever contractor deliberately alters the brief by stating alterations or hedging his bets by introducing a long list of provisional sums, over and above those stated in the employer's requirements? The obvious answer is that the employer or his agent must examine each such alternative and either accept or reject it at the tender stage—or otherwise the contractor's proposals will prevail over the employer's requirements.

Sometimes the employer or his agent have to consider contractor's own proposed provisional sums for the supposedly unpriceable items in the employer's requirements document to ensure that these are compatible with what other tenderers are offering on a fixed price basis. The trick here is to insert knowingly low provisional sums and so create an artificially low tender, but with the opportunity of negotiating cost plus on final account.

The usual set of employer's tender enquiry documents are then completed by the issue of:

- A contract sum analysis—see Chapter 6;
- A form of tender, including required forms of performance bond and warranties.

Finally, in a design and build competitive tender situation reference should be made to the NJCC revised *Code of Procedure for Selective Tendering for Design and Build* which, *inter alia*, suggests a two-stage approach, with unsuccessful tenderers being reimbursed part of their tender costs—much like a limited design competition where the employer guarantees a stated minimum sum of money for each competitor, whether successful or not.

2.4 NEGOTIATED TENDERS

Whereas in a competitive tender situation there is a school of thought amongst contractors which says "If in doubt, leave it out", in a negotiated tender situation the reverse can apply. This is best explained by a cautionary tale from several years back. I was asked at short notice to go down to the West Country and negotiate with a design and build contractor for a very large distribution centre to be built on a green-field site.

The problem was that unless I got the build price down, the three-way development deal between the contractor (who happened to own the preferred site), the employer (a national distribution company) and the end user (a national food retailer) was off. Further, Phase 1 of the contract had to be up and running within 11 months of the herd of cows being driven out of their pasture and Phase 1 unlocked access to further development of the rest of the site.

Working late at night in a local hotel I was struggling with the contractor's sketchy drawings and his approximate quantities. As such, I was making no real inroads into the bottom line build-out price—until I thought of checking the draft development agreement. In that document, the contractor was responsible for all main services throughout the total site, yet his approximate quantities included the main sewer in our Phase 1 package. The next morning the contractor pleaded an "innocent mistake by my estimator" and offered his apologies, which I happily accepted.

The subsequent reduction enabled the three-way deal to go through and the project started almost immediately. In the meantime I visited Saudi Arabia and on one site in Taif we were rock-breaking through "soil" somewhere between granite and marble. When I returned to the West Country distribution centre site three days later and was parking my car I heard the unmistakable sound of four rock-breakers hammering away—and I could not believe I was back in the UK. That sewer cost someone a bundle of money!

It turned out that the cows' pasture consisted of no more than 250 mm of topsoil overlaying solid granite and I subsequently learned that only 250 metres away was the site of the longest running construction claim ever—the Great Western Railway cutting where in *circa* 1840 the contractor made a claim on the proprietors and Mr IK Brunel, the engineer, for unforeseen ground conditions. Seemingly that claim went all the way to the House of Lords and by the time their Lordships' decision was handed down some 70 years later all the original parties were bankrupt and/or long since dead!

So there are these three lessons to be learnt from this cautionary tale:

- Make sure that employer's requirements documents align with any development agreement.
- On a design and build contract make very sure *who* is taking the risk of ground conditions.
- In a dispute situation it is never too early to settle.

CHAPTER 3

THE PHILOSOPHY OF DESIGN AND BUILD

3.1 EMPLOYER'S OBJECTIVES—HOW TO EXPRESS THEM

All clients know what they want—or think they do. However, given the fact that the employer-to-be at project inception stage is usually a collection of individuals and/or outside consultants advising them on their development and procurement options there is wide scope for indecision and lack of clarity in the original brief.

When such matters as budget and planning applications are fed into the equation the original brief can change substantially, but sooner or later the employer and his professional advisers must fix the brief for tender purposes such that comparable competitive tenders can be obtained. Only once an acceptable tender has been examined and approved will a client commit to the project and a contract be entered into between employer and contractor.

The above comments apply to all construction contracts but in terms of design and build the formulation of the employer's brief is all-important.

At least under the JCT 80 Standard Form With Quantities one has a sporting chance of being able to negotiate any post-contract changes of mind by the employer, but under JCT 81 With Contractor's Design and no bill of quantities it can be akin to playing poker, if the tender process is not properly controlled.

Thus where the employer is not prepared either to invest in a design team to work up a detailed design, sufficient to tender on the basis of JCT 80, or needs to save time by getting into early contract, e.g. to qualify for funding grants before the financial year ends, JCT 81 With Contractor's Design might seem an attractive and alternative route.

However, whatever the pressures to sign up with a design and build contractor the employer must have considered the other implications of opting for JCT 81, such as:

- How far to take the design before handing it over to persons unknown?
- How can the employer be sure to obtain the finished product he envisaged but has yet to describe in defined terms?
- What if the employer needs to instruct changes post-contract?

19

- How will costs and timely performance by the contractor be controlled?
- Who in the employer's organisation will act in a personal capacity as employer's agent or will it be necessary to appoint an outside consultant?

All the above factors will bear upon the strategic decisions the employer must make and develop before he can draft the employer's requirements document. Top of the employer's agenda must be the following:

- Statutory consents: e.g. who will obtain outline planning consent and when, if not already obtained?
- Design concept: e.g. totally flexible or preferred building form, shape, height and elevational treatment.
- Essential planning information: e.g. number of occupants, declared function, preferred space planning.
- Site conditions: e.g. restrictions and ground conditions.

Typically the employer will settle for obtaining the outline planning consent himself, as otherwise all future own time and external fee commitment could be in vain. If this is the case the submission to the planning authority will have described the type of building and its intended usage. At the same time the employer will usually have formulated his first thoughts on space planning or on the logistics of his proposed operation.

Accordingly the employer's requirements usually enclose schematic plans and elevations, leaving the tenderers to develop the internal and external details such that once the preferred solution and tender are accepted, a submission can be made for full planning consent—usually by the selected design and build contractor once appointed.

In these days of computer-aided design drafting facilities most employers have access to such a resource, either in-house or out of house, so this is the easy part of the employer's requirements. Thereafter the more difficult part is knowing how far to go in telling contractors what you would like, rather than factually what you must have for your money in performance terms and in describing the finished *quality* required.

Essentially the golden rule is: limit the employer's requirements to necessary matters, i.e. specific contract particulars, functional matters being described by way of a performance specification, including the requirement for the contractor to comply with all current and relevant statutory matters including British Standards, and any preferences as to mechanical and electrical systems.

The question of *quality* is best covered by a schedule of required external finishings, e.g. brick selections (being a planning-reserved matter) and internal finishings described against room types, with other than basic finishings benchmarked by setting PC (prime cost) prices for materials, e.g. ceramic wall tiles PC £45 per square metre.

In recent years employers have tried to have it both ways by fully designing schemes with the usual design team and then going to tender using JCT 81 With Contractor's Design, i.e. requiring all the supposed design and build contractors on the tender list to fully validate the preconceived design and then give the design team, and thereby the employer, absolution from any liability in the project as built. Apart from being very wasteful of resources this is little more than "risk dumping" and this flies in the face of the philosophy, and perceived benefit, of design and build as a procurement strategy.

Whilst the employer is no doubt seeking single-point responsibility he is denying himself the benefit of "buildability"—that elusive skill which contractors are supposed to bring to the table for the benefit of the enlightened employer. Clearly if the employer can bring together the skills of the concept architect and the contractor's planning engineer *before* starting on site he should get the best of both worlds, usually in terms of *time* and therefore *cost*.

The enlightened employer will also be aware that design decisions have not only direct cost implications, but also indirect life-cycle cost implications. However, when up against capital budget restraints it is a bold decision to forego *quantity* for better *quality*.

Unfortunately the recent recession has depleted contractors' and consultants' ranks alike, and the skill in interviewing potential design and build contractors is in identifying those that do retain old-fashioned planning engineers, i.e. production engineers who know how to build any conceived scheme most effectively.

All too often, despite promises to the contrary, one finds contractors who are sadly little more than commercial letter boxes, placing all work, including design, with subcontractors. Such contractors lay off all risks and make their money by creating buying margins combined with non-back-to-back payment terms. To add insult to injury, they then charge the employer for managing little more than paperwork.

A cynical view maybe, but a view now recognised by the more enlightened contractors who are promoting "partnering" and negotiated contracts. There is even a group of leading mechanical and electrical contractors now promising to be "best practice contractors" and who offer to deal with their subcontractors according to a Code of Conduct. The proof of the pudding will be in the eating, as the expression goes.

Following on this last point, one of the abuses which has dented the reputation of design and build is this "subcontract everything" mentality. This often manifests itself in the placement of major works packages, including design, e.g. mechanical and electrical services, with domestic subcontractors who are not known for their design capability.

This is very unsatisfactory for a number of reasons, not least:

- Remoteness of design responsibility;

- Tendency for subcontractors to offer basic solutions, materials and preferred plant (maximum discount) led;
- Risk of insolvency of domestic subcontractors—losing the benefit of any design warranties and PI cover, if given;
- Reactive rather than proactive co-ordination by the contractor of working drawings to avoid design and installation clashes;
- General attitude of "if there is a problem it is either the employer's fault or our subcontractors' ".

It is therefore important that the employer's requirements document lays down the ground rules specifying:

- Who in the contractor's team will lead and deliver the necessary co-ordinated design development for the project;
- If the contractor wishes to subcontract any aspect of the design, that the employer's written consent is first obtained;
- That the employer's professional advisers are allowed to visit the contractor's subcontractor's design offices and satisfy themselves of their capability.

Different projects will require different approaches as to how the employer expresses his objectives, but certain first principles will be common to most, if not all, design and build projects.

The employer's requirements should set out *what* is to be built, but not *how* it is to be built and should be flexible enough to permit alternative design solutions and alternative construction methods, e.g. *in situ* concrete frame or structural steel. Indeed, alternative design solutions should be welcomed as part of the tender enquiry.

As regards practical completion and handover procedures, the employer's requirements should be specific, including:

- Notice period for proposed completion;
- Who will represent the employer at handover inspection;
- Record drawings and manuals;
- Test certificates;
- Commissioning of services;
- Building control sign-off.

The employer's agent who does nothing more than turn up on the appointed day, walks round the site, accepts two boxes of drawings and manuals, signs the contractor's proforma and goes to lunch is taking far too much on trust, and playing into the hands of the design and build contractor.

How can he, an architect or quantity surveyor, know that all the mechanical and electrical systems have been tested, if he has not first insisted that such tests are done in the run up to practical completion and have been witnessed by his designated services engineer on behalf of the employer? How can he know whether the record drawings are right, wrong or indifferent if he has not even unfolded them?

All basic stuff maybe, but sadly it does sometimes happen like that, so get it right at the front end: lay down precise handover and commissioning procedures in the employer's requirements.

Finally, all employers and their agents or solicitors setting up contracts should be familiar with the very useful employer's requirements check-list offered by the JCT Practice Note CD/1A, particularly the long list of contract particulars which need to be put in the tender enquiry, if they are to be subsequently incorporated in a signed JCT 1981 With Contractor's Design Contract.

Unless these requirements are spelt out as part of the tender process there may well be problems in getting them incorporated by agreement in the signed contract, or worst of all, important details might be missed altogether, only to emerge as problems during the project.

So the bottom line message to employers and their agents has to be: take that little bit of extra time in preparing the employer's requirements document. If possible get an independent mind to troubleshoot it, with the brief "Are the employer's stated objectives clear and complete, sufficient for tender purposes?"

3.2 CONTRACTOR'S OBLIGATIONS AND RISKS

The design and build contractor's overriding obligation is to comply fully with the employer's requirements, but problems will inevitably arise where there is any perceived lack of clarity, ambiguities or obvious gaps in the tender enquiry.

Assuming there is a competitive tender situation and there is some doubt as to any matter in the employer's requirements, the design and build contractor has essentially three choices:

- Write to the employer seeking clarification of his requirements during the tender period;
- Make a reasoned decision as to what might be required and state the assumption in the contractor's proposals document;
- Take a "minimal" view and only price what is definitely required, drawing the employer's attention to any "grey" areas by way of stated qualifications or exclusions.

Any other course of action, I would suggest, is unsatisfactory in that Clause 2.4 of JCT 81, Discrepancies within documents, provides that when the contractor's proposals have not dealt with any discrepancy within the employer's requirements, the contractor shall submit his preferred solution in writing to the employer, who can then adopt such solution or propose his own. In either case the agreed solution is deemed to be a Clause 12 change, so inevitably there is plenty of scope for disagreement, particularly if the costs are significant.

In these circumstances, I would suggest it is far better for all concerned if the design and build contractor notifies the perceived discrepancy or other uncertainty, preferably by a positive statement in the contractor's proposals, rather than keeping quiet in advance of tender acceptance and supposedly discovering the problem when the employer is contractually vulnerable, which amounts to a prepared "ambush".

For the above reason I like to see the contractor's proposal document structured paragraph-by-paragraph to mirror the employer's requirements and offering objective comment, or confirmation as applicable. In Chapter 8 the JCT 81 provisions of Clause 2.4, Discrepancies within documents, are further discussed, including discrepancies *between* documents.

As regards genuine uncertainties, I see no objection to the design and build contractor offering provisional sums for undefined items subject to three caveats. These are that:

- Such provisional sums are sensible values;
- There are not too many of them;
- The tendered design fees are inclusive of the work involved in procuring such items.

Otherwise the contractor's obligations can be categorised under two headings:

- Design development
- Build-out

Under JCT 81 design development involves taking over *full responsibility for all design matters from the point reached by the employer's requirements.* This responsibility includes:

- " . . . the selection of any specification for any kinds and standards of the materials and goods and workmanship to be used in the construction of the works so far as not described in the Employer's Requirements or Contractor's Proposals"—Clause 2.1 extract.
- " . . . any further design which the Contractor is to carry out as a result of a Change in the Employer's Requirements . . . "—Clause 2.5.1 extract.

Design development will firstly include the obtaining of any outstanding and necessary statutory consents such as Building Regulations approval.

Secondly it will include the submission to the employer's agent of all samples of materials or details of specialist systems which may have been expressly reserved for the employer's approval in the employer's requirements, and obtaining the employer's written consent thereto.

Sensibly design development will also include:

- The submission of all working drawings and main plant manufacturers' catalogues to the employer's agent such that the employer, through his agent, is given the opportunity to comment on, or even to veto the contractor's detailed design and systems selection;
- The submission to the employer's agent of any alternative design solutions or materials to those which have previously been identified in the employer's requirements or contractor's proposals;
- The early submission and agreement with the employer's agent of sample boards of finishings, e.g. carpet, tiles, wallpaper, etc.;
- The early completion of a sample room to establish the standards expected in joinery, plastering, etc., as well as sample doors and windows.

The Design and Build Contractor therefore takes on not only the normal range of risks taken by a JCT 80 Contractor but he also takes obligations and attendant the risk of:

- Design development generally;
- Clearance of all outstanding planning and building control matters;
- Timely production of approved working drawings;
- Co-ordination of all services detailing and associated builder's work;
- Production of record drawings;
- Incorporation of the employer's reasonable requests for changes, so far as they affect the design;
- The possibility of technical or commercial failure by design consultants or subcontractors with design input;
- The possibility of latent and patent defects in the design.

To guard against the commercial failure, i.e. insolvency, of a subcontracted design consultancy or domestic subcontractor offering design of specialist works, the design and build contractor will usually have given the employer the benefit of professional indemnity cover to the value agreed in his own name, but in real terms what happens if, say, the electrical subcontractor goes into receivership six weeks short of practical completion?

The immediate answer is normally that the design and build contractor will re-employ the key operatives and take physical possession of all electrical materials, locking them away before the receiver's agent or the suppliers come looking for them.

The ex-directors of the insolvent company have a duty to assist the receiver in general debt collection etc., so they will not normally be available to the design and build contractor, and usually the contractor will be the last person they will wish to help, especially if non-payment has allegedly caused or contributed to their misfortune.

As regards the actual work done, and yet to be done, there may be temporary labelling of partly installed and untested electrical circuits and no record drawings. There may also be outstanding orders for key components such as the fire alarm installation, all within the electrical works package.

Not unnaturally third party subcontractors and suppliers with invoices outstanding will require new subcontracts and advance payment, so the design and build contractor has to dig deep in his own pocket to keep the show on the road—but what about continuity of design and the contingent liability of design error, e.g. undersizing of plant or cabling errors?

These are very real risks for the design and build contractor and really he has little option but to bring in an independent consultant to completely revalidate the design and prove all existing circuits etc. It may even be necessary to change fundamentals such as fire strategy, e.g. fusible links on corridor smoke doors—all with just six weeks to go to practical completion, not to mention the further risk of liquidated and ascertained damages being applied by the employer.

In extreme cases of subcontractor non-payment, the design and build contractor may even have to answer to the receiver or liquidator under the Insolvency Act 1986 and face a claim for the full costs of the insolvency, but then he probably only has himself to blame if the position is that extreme.

The insolvent electrical subcontractor will probably have given a warranty in favour of the employer, but this is of no value to the design and build contractor, warranties only being useful where there is a three-party situation, A to B to C, and where the middle party, party B, ceases to trade. In the above case, it is party C, the electrical subcontractor that has disappeared, leaving party B, the design and build contractor, totally exposed to deliver his full obligations to party A, the employer.

Any design fault discovered *before* the insolvency of the electrical subcontractor will be the responsibility of the electrical subcontractor's professional indemnity insurer, who gives the design and build contractor protection provided the claim has been notified prior to receivership and provided he has not constrained the design role of the subcontractor. In other words, if the contractor has dictated design revisions or downgraded plant and material specifications to create cost savings contrary to the subcontractor's recommendation, he may well not be covered by insurance.

However, the real risk is where the design and build contractor takes a chance and does not revalidate the design of the insolvent electrical subcontractor prior to practical completion. Professional indemnity claims are only met on a "Claims made" basis, i.e. when discovered and reported to insurers—not when the design error originated.

Thus the design and build contractor will again be on his own if the electrical system starts tripping out after practical completion and no run-off professional indemnity insurance has been paid for. There is simply no point in suing receivers!

So the underlying message for design and build contractors is: look after your subcontractors, particularly those with significant design input.

3.3 CERTAINTY OF COST AND TIME

In its simplest form design and build can be a total "turnkey" project, i.e. the contractor takes the employer's brief, agrees the price and time, negotiates any construction problems, and delivers the project by handing the employer the proverbial keys to the door.

This concept is more frequently applicable in the case of overseas manufacturing complexes where the employer is geographically far removed from the site, but it requires a high degree of trust both ways, with or without third-party audit visits.

On such a project there will usually need to be certainty of *cost* and *time*—as otherwise it would not have been commissioned—and a robust, pragmatic view will be taken of any problems as and when they arise. Only if necessary will a new price and new end date be agreed. In practice it may not be possible on overseas projects to resource materials in the event of proposed changes without delays, because of customs and import controls, and these factors may well limit the opportunity for changes.

In theory, design and build contracting in the UK should be no different, except that two factors tend to militate against the "turnkey" approach.

- The proximity of employers to "their" site and the natural temptation to keep refining their requirements;
- The muddled attitudes of contractors and in particular poorly trained site staff.

Typically a UK contracting organisation will have different operating divisions—one offering lump sum contracting (JCT 80) and one offering design and build (JCT 81). In some cases the same division will offer both forms of procurement. In these circumstances it is hardly surprising that at middle management and site level the essential differences between JCT 80 and JCT 81 slip out of sight, and standard JCT 80 systems and proformas are used for JCT 81 design and build projects. How often do we see the same attitudes and correspondence? It is always someone else's fault when things have started to run late, the finger usually being pointed at the architect/employer's agent for late release of supposedly requested information, yet on design and build it is the contractor who should be in control of the information flow. A claim for loss and expense is rarely far behind.

Happily there are several design and build contractors who fully subscribe to the different philosophy of JCT 81, take care to train new staff who come from JCT 80 contracting, and have specific procedures and proformas geared to JCT 81.

However, the problem repeats itself at the specialist subcontracting level, in that if the design and build contractor subcontracts the total package, i.e. the work and the design, he may well do so to a specialist firm which works under both systems, JCT 80 and JCT 81, and on whom the subtleties of JCT 81 are lost! As far as such a firm is concerned, a variation is a variation, and the debate under JCT 81 as to whether it is design development or a Clause 12 change is academic!

The design and build contractor's site quantity surveyor could therefore find himself with a non-back-to-back situation and must be clear how to handle it—even if it is to his own employer's disadvantage.

Certainty of time *and* cost *is therefore an ideal rarely achieved in UK design and build practice,* unless *specific provisions and procedures are written into the tender enquiry and the subsequent contract.* Even then, firm post-contract supervision by the employer's agent will be needed to deal fairly with contended changes to the employer's requirements, and the almost inevitable *time* and *cost* implications.

If the project is properly handled by the employer's agent and the contractor's middle management there is no reason why there should not be a forward cost control system and therefore a rolling final account. Likewise, any time implications should be negotiated on their merits.

In the early 1980s I had such a project under the standard JCT 81 Form of Contract, as it then existed, and at the first site meeting we agreed that if the design and build contractor wished to raise any change orders arising out of the employer's approval or comments made on the contractor's working drawings then he had to raise a change order request proforma, which both parties had to agree and sign before the work could be carried out.

A broad brush approach was taken and certain refinements were pre-agreed as regards costs, but the end date remained unchanged. All was sweetness and light until the first lorry arrived—full of dog food as it happened—but that is another cautionary tale: see Chapter 8.

At about that time the British Property Federation were becoming concerned that there seemed to be no more certainty in the two key areas of *cost* and *time* under design and build than under the more traditional architect-designed lump sum contracting of JCT 80, so the BPF persuaded the JCT Committee to formalise, by way of optional inclusion in the JCT 81 With Contractor's Design Form of Contract, what some practitioners were already doing, i.e. requiring pre-agreement of contended Clause 12 changes in respect of both *cost* and *time*.

Thus in February 1988 by way of Amendment No. 2 the optional "Supplementary Provisions S.1 to S.7" were introduced.

So we now have a choice between a "soft" JCT 81 contract—without the Supplementary Provisions and where the *cost* and *time* effects of changes are at large—or a "hard" JCT 81 contract—with the Supplementary Provisions applying, which if properly observed will give certainty of *cost* and *time*.

These optional Supplementary Provisions, covering Clauses S.1 to S.7, are not to be confused with the Supplemental Provisions of JCT 81 covering VAT and are so important that they are separately discussed at Chapter 5.

3.4   THE ROLE OF CONSULTANTS AND NOVATION

As has been discussed in Chapter 2 the essential distinction of JCT 81 is that the detailed design is the responsibility of the contractor, not the employer, and that accordingly the post-contract architect and the supporting design team are employed by the contractor, either in-house or out of house by way of a consultancy agreement.

In the event that the successful design and build tenderer is one of those that retain in-house design teams or will only use one or two proven outside consultancy practices it follows that the architect engaged in the first instance by the employer to develop a scheme concept will have a limited role, i.e. pre-tender only. The same applies to the other professionals who may have been called in by the employer at the early stage: usually a quantity surveyor to prepare a cost plan, and sometimes a structural engineer, particularly if a soil survey is to be carried out.

As and when the project proceeds to contract award the employer's selected design team will have finished their task and will therefore be disbanded, unless:

- The employer, as part of the tender enquiry, has required that the contractor will re-employ the pre-contract design team—"Novation" is then incorporated in the signed contract; or
- The employer re-engages one or more of his original design team to fill the role of employer's agent—and the contractor appoints his own separate design team.

"Novation" is a legal concept, formally providing for the re-employment of a consultant by the contractor, as part of the contract agreement between employer and contractor. Should the contractor merely choose to re-employ the same individuals who have previously sat the other side of the interview table from himself that would not be novation, but its effect would be very similar.

The most obvious problem with novation is the overlap period, i.e. the period when the architect and others are still employed by the employer during the tender period, yet the contractor-to-be is appraising the employer's requirements and supposedly providing lateral thinking as to design solutions and buildability. Obviously in a competitive tender situation it would be entirely inappropriate for the architect at that stage to have any direct dealing with any of the tenderers, but in a negotiated tender situation there is more opportunity for co-operation. The architect will however still have to report to his client, the employer, and provide objective advice, even if it might prejudice his prospective re-employment.

It is therefore difficult to see how the employer, in a competitive tender situation, can expect any lateral design thinking from tenderers if the tender enquiry requires novation or "suggests" the successful tenderer might "like to consider re-employing" the employer's preferred architect or other members of the original design team.

As against the perceived benefit of design continuity, the employer requiring novation is denying himself the value engineering benefit of an independent professional, probably with a contracting background, making a critical analysis of the employer's stated design concept. Inevitably two or more minds may be better than one, particularly if blinkers are being worn by the one.

Moving on therefore: once the contract has been awarded and novation of the design team has occurred, what are the problems post-contract?

The immediate risk for the employer might be described as a tendency to "take the eye off the ball". In the euphoria following completion of a possibly drawn-out or difficult tender selection process the natural reaction is to catch up with other pressing priorities, which may have been set aside, and to assume that the design and build contractor will now take the reins.

If so, the honeymoon period is likely to end with a rude awakening when the employer is required to give site possession, or even to pay the first valuation. Despite promises from the contractor, key documentation may still not be available, e.g. the performance bond, the signed warranties, the contractor's programme, the form of contract itself, the contract sum analysis back-up, etc.

A properly organised employer will have appointed an employer's agent to fill the void left by the novated design team and will have set up procedures and a tight timetable for completion of all documentation before giving site possession and *certainly before being liable to pay the first valuation.*

The novated architect and his fellow professionals will now have suddenly found themselves in a strange environment—much like professional footballers who find themselves transferred to another club and then rubbing shoulders with new colleagues whom they were kicking only the week before.

Inevitably the architect's services will be required immediately by the contractor to advance the design development and obtain any necessary statutory consents, but maybe the novated design team's exact terms of appointment and duties will have yet to be resolved with their new employer, the contractor.

So long as the novated design team are needed by the design and build contractor their telephone will ring, but once the second-phase design development drawn information has been completed they may well find themselves relegated to the second XI, only being brought out for the occasional first XI fixture, so to speak.

The contractor's buyer and quantity surveyor will increasingly be calling the shots and until the project gets out of the ground there is very little site supervision for the novated architect to do anyway.

The novated structural engineer is likely to find himself similarly used to validate initial designs and calculations, but thereafter he is likely to be consulted only when there is a particular problem, as the contractor's site team is likely to be led by trusted staff engineers.

The original quantity surveyor really has nothing to offer to a design and build contractor, particularly as he will be unfamiliar with the contractor's in-house buying and subcontracting practice. In fact, there is probably very good commercial reason not to take on the employer's quantity surveyor. However, the employer still needs someone to look after financial issues arising under the contract and it is more likely the quantity surveyor will find re-employment, either as the employer's agent, or as a consultant advising whoever the employer choses to appoint as the new "architect", i.e. employer's agent.

Equally, if there was a services engineer on the original employer's design team it is doubtful whether he has much to offer the design and build contractor, unless the contractor has the good sense to appreciate the real worth of a professional services co-ordinator, and does not have such an individual already on his staff.

Thus the tendency on design and build contracts where novation of consultants is involved is for the employer to gain a false sense of security, believing that the trusted individuals transferred to the contractor will be retained as first XI regulars—so providing continuity of design and offering full post-contract supervision, such that a fully compliant building is delivered, with no unauthorised short-cuts or material substitutions. Such naivety can have a heavy price tag!

Any novation agreement set up between employer and contractor should therefore provide for:

- Full disclosure by the contractor to the employer of the terms of appointment offered and signed by the professionals including specific descriptions of duties to be performed;
- An undertaking that no professional appointment by the contractor will be changed in any way without the prior knowledge of the employer;
- The production by the contractor of his own professional indemnity policy to show the scope and limit of cover, together with proof of current premium payment;
- Similar production of professional indemnity cover for *each* of the novated professionals;
- A commitment to maintain run-off professional indemnity cover for 6 years, or 12 years if under seal, and to produce proof of payment of all future premiums.

It follows that once novation has taken place the employer must not delay in enforcing the above, having first written in a clause in the employer's requirements to the effect that there shall be no valuation payment until such documentation is satisfactorily produced by the contractor.

It is not unknown for defects to be discovered after design and build contractors have become insolvent, in which event the employer will then look to the PI policies of the novated professionals. On one such contract it was then found that the contractor had retained the novated professionals for design development only; they had no duty of supervision, so the architect's PI policy, which included supervision of projects where so required, did not bite on the project in question, leaving a £5m headache for the funder when the employer/developer with incestuous links to the failed contractor promptly folded.

Surprise, surprise—it then turned out that the employer's agent was not an independent professional, but a limited company closely related to the failed developer, so the finger of blame went full circle back to the funder's solicitors and the separate quantity surveying practice who had been appointed as project monitors.

It is therefore essential that the design and *supervision jigsaw of professional responsibilities is complete*, and care must be taken to ensure that in the event of the disappearance of any one party there is alternative linkage with no discrepancy between terms of appointment and terms of PI cover.

Again, it must be stated up front in the employer's requirements that any attempt by the contractor or any member of his chosen design team to hide behind the supposed confidentiality clause of their PI insurance policies will not be acceptable—nothing less than full disclosure of insurance cover to the employer on a confidential basis should be a condition of tender.

In the case of novated design teams, it is assumed that the employer will have satisfied himself as to their PI insurance cover before appointing them in the first place, but this could be an unfortunate assumption. As part of the novation process all such policies and proof of current cover should be rechecked.

Novation is therefore a difficult area for various reasons and if contemplated by the employer, for whatever reason, needs very close legal advice and attention to make sure there are no missing pieces in the jigsaw of responsibilities. Wrongly handled novation can create worse problems than it is designed to avoid.

Although I have not been involved with a project on which "double novation" has applied, I understand this is becoming more popular with supposedly enlightened employers.

Basically the concept of double novation is that the employer selects his design team, develops his requirements to whatever level of detail he considers appropriate and then tenders the project on condition that:

- The contractor will re-employ the design team upon contract award, *but only up to practical completion*.
- Thereafter the design and build contractor's individual contracts with each member of the design team are automatically determined *and they are then re-employed by the employer*!

Presumably the employer will have retained a no-nonsense project manager throughout as the employer's agent—but can this set-up really work in practice? How can an architect or engineer suddenly become the judge in his own court concerning any alleged design deficiency or compliance issue? How does he square his responsibilities with his PI insurer? The jury, so to speak, is out, and I await the first claim with interest.

3.5   THE EMPLOYER'S AGENT

The concept and legal entity of the employer's agent is unique to design and build. Essentially an employer's agent is the technical and professional ears, eyes and mouthpiece of the employer, who will often be a lay client or a developer organisation which has no suitably skilled and available in-house manager capable of filling the role.

In fact, it is often better if there is some "distance" between the employer's executive and the professional contract manager as sometimes difficult decisions need to be taken. Arguably there is more chance of objectivity, if matters can in the first instance be considered on their technical and contractual merits, other than by the person directly responsible for the budget or profit and loss account.

In the case of JCT 81, Practice Note CD/1B spells out in the Commentary on Article 3 the key features of the role of the employer's agent:

- "The Employer's Agent is referred to expressly only in Clause 5.4 and in Clause 11 under which the Contractor is obliged to afford the Employer's Agent and any person authorised by the Employer or by the Employer's Agent to have access to the site and workshops."
- "With those two exceptions the contract treats his acts as the acts of the Employer."
- "The naming of a single person as Employer's Agent does not preclude the employment of other persons by the Employer to advise him or the Agent for the purposes indicated in paragraph 4 of Practice Note CD/1A."
- "Unless the Employer by written notice in accordance with Article 3 informs the Contractor otherwise, the Contractor can and must regard the Employer's Agent as the duly authorised agent of the Employer for the performance of any of the actions of the Employer under the Conditions."
- "Whether or not the person so named as Agent is an Architect or a Quantity Surveyor he acts as the Employer, and not as certifier or valuer between the Employer and the Contractor."
- "The Employer is entitled to remove and appoint his Agent at will, but must inform the Contractor of his Agent's identity."

Before we consider where all this leaves the contractor in terms of getting paid and having the work approved, some further comment is called for in respect of these six basic provisions defining the role of the employer's agent.

Firstly, the JCT 81 contract is careful to be consistent, i.e. everything the agent does is deemed to be with the full knowledge and consent of the employer, even though the agent may well make decisions in good faith without reference back to his client. Thus throughout the contract, obligations owed to the contractor are owed by the "Employer", not by the "Employer's Agent".

Practice Note CD/1A refers to four purposes for which the employer may wish to rely on his own architect and/or quantity surveyor. The first three are all pre-contract:

- "Formulate the Employer's Requirements."
- "Make a preliminary appraisal of what Contractors have to offer so as to advise the Employer in selecting Contractors to be invited to tender."
- "Make an appraisal of tenders to enable the Employer to make an informed choice."

The only post-contract purpose suggested by Practice Note CD/1A is wide in the extreme and is in effect a monitoring or audit role reporting direct to the employer or as a sub-consultant to the appointed employer's agent.

- "Advise the Employer or the Employer's Agent on the Contractor's execution of the Works and payment therefore."

Article 3 of JCT provides for the naming of an individual to act as employer's agent:

" . . . or such other person as the Employer shall nominate in his place for the purpose shall be the Employer's Agent referred to in Clause 5.4 and 11 and, save to the extent which the Employer may otherwise specify by written notice to the Contractor, for the receiving or issuing of such applications, instructions, notices, requests, or statements or for otherwise acting for the Employer under any other of the Conditions."

At least Article 3 provides "in writing", and sensibly this written requirement must apply in the circumstances where the employer wishes to disinstruct or otherwise change his agent. Clearly the contractor must be told, so he can react accordingly, knowing who has or has no longer got authority on behalf of the Employer. A purely verbal instruction on such an important issue could cause all sorts of problems, or be abused by a contractor wanting to create mischief.

In practice employers do for various reasons wish to change their agent. Such reasons can be:

- Completion of one phase of the works with a different technical emphasis on the following phase;
- Incapacity of the original employer's agent;
- Loss of confidence or a dispute with the original employer's agent;
- Resignation of the employer's agent;
- Clash of personalities with the contractor.

The important point is that whilst the original appointment of the employer's agent is at the absolute discretion of the employer, like the employment of the architect under JCT 80, that discretion is extended under JCT 81 to changing the individual mid-project.

The concept of changing the employer's agent mid-project is a significant departure from JCT 80 and goes to emphasise that under JCT 81 the employer's agent is the employer's man, with no third-party umpire function or duty of care to the contractor. Further, no reason has to be given to the contractor when effecting a change of employer's agent and, in JCT 81 practice, change can also mean the splitting of duties.

A good example would be where the employer might wish to retain the original employer's agent for continuity of design approvals, specification compliance and measured work final account agreement, but the contractor has made an extension of time claim or loss and expense claim which directly or indirectly points the finger at the original employer's agent. In such circumstances the employer might wish to be separately advised and try to show objectivity to the contractor in having a fresh mind to consider the facts.

Another example would be where defects are an issue and the employer has to balance the need to take possession versus what he can accept as remedial works whilst in occupation, or choose to live with subject to agreeing a discounted value. Given this scenario the employer may well opt to appoint one of his premises managers as employer's agent to deal with remedial works, but he must be very careful to define the role and confirm same in writing to all interested parties.

So we now jump the fence and ask: How does the design and build contractor see life under the JCT 81 regime, with no third-party umpire? Being cynical, one might answer: no differently than under JCT 80, as although architects, quantity surveyors, etc. are given third-party umpire status under the terms of the JCT 80 contract when all is said and done, their fees are paid by the employer and he who pays the piper calls the tune.

Happily the majority of professionals acting under JCT 80 still respect the special status given to them as third-party umpires and try to be impartial in the difficult decisions. Equally most contractors accept that professionals wear two hats and, on the face of the contract correspondence at least, accord them the benefit of doubt in their secondary role of third-party umpire.

JCT 81 however does away with the Chinese wall which metaphorically divides the two roles of the JCT 80 architect. Under JCT 81 the employer and his agent are on one side and the contractor and his team, i.e. architect, engineer, etc. are on the other, with the contractor being deemed sufficiently worldly-wise as to be able to fully represent himself in all contractual and financial matters.

The scene is therefore set under the basic JCT 81 provisions for a head-to-head clash should the project not go according to either the employer's or the contractor's anticipations.

The design and build contractor, once appointed, is in total control—or should be—and apart from checking and paying out money when the contractor submits an invoice the employer and his agent are reduced effectively to the role of spectators, but with the right of veto, holding up a yellow or red card as appropriate.

If the design and build contractor wishes, he can make wholesale changes in terms of material selection, procurement and construction sequencing, taking the benefits to himself which he should have put on the table at time of tender. If, however, such wholesale changes compromise the integrity of the agreed design concept or any express provisions within the employer's requirements or contractor's proposals it is up to the employer's agent to spot the differences, hopefully at drawing approval stage, or at least no later than when such unauthorised changes appear on site. If the employer's agent fails to spot the differences or only blows the whistle at practical completion, or thereabouts, it puts employer and contractor on a collision course.

Thus the "thinner" the employer's performance specification, the greater scope there is for the less scrupulous design and build contractor to substitute alternatives as against what was allowed in his tender and to make hidden savings. Conversely the fuller the employer's performance specification the less room to manoeuvre there is for the design and build contractor, but that then denies the intended flexibility of design and build.

In subsequent chapters there is full discussion of what happens when one party perceives that the other is not performing his side of the bargain, i.e. *cost* or payment issues (Chapter 6), *time* issues (Chapter 7) and *quality* issues (Chapter 8).

However, when it comes to disputes between the employer and the contractor under JCT 81 there is either a single- or two-stage contractual procedure, depending on whether the option of the Supplementary Provisions have been incorporated in the contract. If the JCT 81 Supplementary Provisions are incorporated the basic "soft" contract becomes a "hard" contract, with a strict procedure for forward agreement of Clause 12 changes and very real sanction for the contractor who does not so comply. Commentary on these unique and radical provisions is to be found in Chapter 5.

The big question regarding the role of the employer's agent is: How should he conduct himself during the contract? Obviously he *must* be careful to:

- Define his authority and lines of reporting, including frequency, with his client, the employer;
- Ensure that such authority, including any changes in the course of the project, are communicated to the contractor in writing;
- Make sure he does not exceed his authority in instructing changes, e.g. funding limits;
- Not leave himself open to the employer claiming he has not been kept informed of any decision given to the contractor concerning *cost*, *time* or reserved *quality* matters;
- Not leave himself open to the contractor claiming delay in approval of reserved matters.

On the other hand the employer's agent must not allow the employer to be too "hands-on" as this will inevitably lead to regular and excessive changes of mind or indecision, which could well give contractual grounds for additional *costs* and *time*.

As regards his dealings with the contractor, the employer's agent, not being a third-party umpire, can take the purist view, i.e. hard line/confrontational stance when there is an issue. However, the commonsense approach is to try to play the honest broker role—holding the ring between employer and contractor and giving technical and contractual decisions no differently than if he were the architect under JCT 80.

It does not serve the employer well if an employer's agent under JCT 81, or architect or quantity surveyor under JCT 80, causes the contractor to adopt a "claims" attitude from early in a project, but on the other hand if the contractor and/or his site quantity surveyor start playing "games" at the employer's expense the employer's agent has to sit hard on the abuse, and the earlier the better.

This is never more true than under JCT 81 and with a design and build contractor intent on cutting corners from time of contract award. The employer's agent must deal with such abuses firmly and should not be shy of ensuring that the contractor's senior management are made aware of any repeated misdemeanours.

Some contractors, including design and build contractors, operate each site as individual cost centres and if faced with such a situation the employer's agent should be particularly alert. Quite possibly the site agent and quantity surveyor are trying to prove themselves "good boys" in the eyes of senior management and so drive their own bargains with subcontractors to maximise profits to the detriment of the *quality* of the finished project and to the employer's disadvantage. Quite possibly they are also creating their own "reserves" with or without the knowledge of head office, and by use of credit notes can confuse the audit trail, and VAT liabilities at the same time.

Certainly such "games" can be played under JCT 80, but without full bills of quantities and with the overlapping of measured work and design costs under JCT 81 the temptation must be that much greater.

And a final observation—the alert reader will have noted that the JCT 1981 Practice Note CD/1B Commentary on Article 3, fourth item, tells the contractor he: " . . . must regard the Employer's Agent as the duly authorised agent of the Employer for the performance of any of the actions of the Employer under the Conditions."

As one of the employer's obligations is to actually pay the contractor it might be sensible to provide in the employer's requirements that the employer's agent in no way acts as fund-holder for the employer and cannot be held responsible for the present or future financial status of the employer. In the event that the employer does become insolvent during the contract it might then be interesting if the contractor runs an argument to the effect that he has been caused loss by the unauthorised actions of the employer's agent, e.g. the issuing of a Clause 12 change instruction causing a cost overrun!

CHAPTER 4

JCT 81 KEY CLAUSES

4.1 CURRENT AMENDMENTS

This book was intended to be different, i.e. not just a clause-by-clause recital and commentary on the most widely used standard form of design and build contract, but a radical analysis peppered with real-life anecdotes to give it some spice and make the reader think "What if . . . ?".

However, given that JCT 81 is the most commonly used standard form of contract used in design and build procurement, being much copied and amended for good reason, it is important that the student reader understands the structure of the contract and that, like a jigsaw, one part depends upon its neighbours. At the same time all readers may benefit from practical comments on the key clauses.

We therefore start our commentary, illogically, with the last page inside the back cover! This is where the current amendments are listed, and the reason for starting here is that the JCT are constantly reviewing and updating all standard contracts in the light of case and statute law and general developments in the industry.

The latest current amendment as at July 1997 is Amendment No. 11 issued May 1997 but it is inevitable there will soon be Amendment No. 12 covering some of the principles of the Latham Report and the Housing Grants, Construction and Regeneration Act 1996.

So, when considering using JCT 81 it is important that in the tender documentation the current amendment number is stated and that when the documents are prepared for signature, after the award of the contract, that the same amendment is incorporated.

4.2 ARTICLES OF AGREEMENT AND RECITALS

Even before the Articles of Agreement, where the employer and contractor are named and the contract is dated, the JCT spell out a warning to would-be "risk dumpers" i.e.:

"This Form of Contract is not suitable for use where the Employer has appointed an Architect/a Contract Administrator to prepare or have prepared drawings, specifications and bills of quantities and to exercise during the contract the functions ascribed to the Architect/the Contract Administrator in the JCT Form With Quantities, 1980 edition (as amended) . . . "

The JCT warning continues by drawing attention to a common situation, i.e. where a certain element of the works is critical to the whole and the design cannot be finalised without input from the specialist subcontractor, who may or may not be known at the time tenders are sought for the appointment of the main contractor. Again, it would be wrong to ask all main contractors to revalidate probably 80 per cent of the detailed design and then to expect the winning tenderer to take responsibility for months of work by the employer's design team, and the JCT 81 contract specifically advises:

"...Where only a portion of the works is to be designed by the Contractor, the JCT 'Contractor's Designed Portion Supplement' reprinted May 1988, for use in conjunction with the JCT Standard Form 1980 Edition (as amended) with Quantities, is available (see JCT Practice Note CD/2)."

The Articles of Agreement then follow, providing for:

- Date of contract
- Employer's name and address of registered office
- Contractor's name and address of registered office

These formalities are common to all JCT contracts and provide for the necessary definition of the usual place of business of the parties such that in the event of formal correspondence or notices needing to be served each knows where they can rely on the other receiving same. In the past "games" have been played, with one party deliberately sending contractual correspondence, which sets the clock running for the other party's response, to other than registered offices at such inconvenient times as 5.00 p.m. on the last normal day before Christmas.

To guard against such "games" it is advisable that where the contract refers to "days" for the purpose of notices these are defined in the "Contract Particulars" section of the specification as "normal working days, Monday to Friday, *not* calendar days". Otherwise, where the contract refers to "days" without further definition, it is to be assumed they are calendar days.

The First Recital provides for a short description of "the Works", the location where "the Works" are to be carried out and introduces the document to be known as "the Employer's Requirements". It follows that the entries to be made are comprehensive and factual; it is no good a contractor signing a contract which describes "the Works" as including mechanical and electrical services and then post-contract trying to plead a mistake in that he thought they were outside his scope of work and to be done by a third-party contractor in parallel with his own work—strange but true!

The Second and Third Recitals are the business clauses recording the events leading up to formal contract signature and introducing the basic offer and acceptance statement:

- "The Contractor has submitted proposals for carrying out the Works (hereinafter referred to as the Contractor's Proposals) which include the statement of the sum which he will require for carrying out that which is necessary for completing all the works in accordance with the Conditions, and has also submitted an analysis of that sum (hereinafter referred to as 'the Contract Sum Analysis') which is annexed to the Contractor's Proposals."
- "The Employer has examined the Contractor's Proposals and the Contract Sum Analysis and, subject to the Conditions and, where applicable, the Supplementary Provisions hereinafter contained, is satisfied that they appear to meet the Employer's Requirements."

The Fourth, and last, Recital deals with the tax status of the employer.

4.3 CONDITIONS

The Conditions of the JCT 81 Standard Form of Building Contract are a first cousin of the JCT 80 provisions in both logic and structure. Where the basic JCT 81 Conditions differ are essentially in five areas:

- Contractors's obligations—Clause 2;
- Employer's instructions—Clause 4;
- Changes in the employer's requirements—Clause 12;
- Payments—Clause 30;
- The optional Supplementary Provisions—Clauses S1 to S7.

The first two have been discussed already at some length in Chapter 3 as they deal with how a design and build contract is set up and how it is the contractor, rather than the employer, who assumes full responsibility post-contract for the design development and final delivery of a fully compliant building.

4.4 APPENDICES

Before we get into the detail of the post-contract key clauses in JCT 81 it is worth stressing the importance of the correct completion of the Appendices section.

This is where the required contract particulars are entered, having, hopefully, first been spelt out to all tenderers in the employer's requirements document. Under JCT 81 there are three Appendices:

- Appendix 1—General particulars ranging from time periods to insurance details etc.
- Appendix 2—Alternative Methods of Payment—see Chapter 6.

- Appendix 3—Provision for the detailed listing of which documents are agreed as comprising respectively the employer's requirements, the contractor's proposals and the contract sum analysis—all as referred to in Article 4.

The Appendix 1 entries as required are again recognisable from JCT 80, but with two notable exceptions:

- The Supplementary Provisions S1 to S7, requiring the option "to apply/not to apply" to be effected by deletion of *one or the other* and the naming of an adjudicator.
- Limit of contractor's liability for loss of use, clause 2.5.3, where the employer has the option of requiring open liability for the contractor should a major design defect cause consequential loss, or the option of stating a financial limit.

In real terms the latter issue will probably be determined by the level of insurance held by the contractor in respect of the specific risk of acting as designer, or that of the private practice to whom the contractor subcontracts the design responsibility. It is therefore important to consider the level required and to state that level in the employer's requirements, thereby putting any tenderer who has not got sufficient cover on notice that he must be able to obtain top-up professional indemnity cover should his tender be success-ful.

The optional Supplementary Provisions S1 to S7 are a subject in their own right and are addressed as such in Chapter 5.

4.5   CHANGES v. DESIGN DEVELOPMENT

Under JCT 80 any changes of mind in the employer's stated design at any stage of the project are his responsibility and are known as "Variations".

Under JCT 81 the same applies, but with two essential differences:

- As the employer's stated design is only conceptual, the contractor then has a major task in developing the design into working drawings and fully co-ordinated "for construction" detail.
- Only employer-required changes of mind, as opposed to design development and fine-tuning by the contractor, qualify as Clause 12 changes.

Often substantial difficulty arises in distinguishing between the two, where in good faith the contractor might make a proposal involving a functional improvement or a material selection. The employer will probably see this as no more than gratuitous design development and approve or even positively comment, maybe adding some refinement of his own. The contractor on the other hand will then pass the employer's approval back to his design office and in due course to his buying department. A month or so later, the site quantity surveyor may well then sit up and claim a Clause 12 change. Who is right?

Obviously the devil is in the detail but if certain basic ground rules are followed there should be no misunderstandings, irrespective of whether the optional Supplementary Provisions are to apply. As such, a robust condition should be written up front in the employer's requirements document to the effect that:

1. The contractor is fully responsible for all design development, material selection and workmanship to a standard not less than that prescribed.
2. The contractor shall submit all working drawings and samples of main materials to the employer's agent in sufficient time as to allow 10 working days for approval.
3. Any approval or other comment by the employer or agent shall in no way constitute a change under Clause 12 unless the contractor shall have first advised the employer's agent in writing of any cost or time implications and obtained the employer's agent own written consent thereto.
4. Clause 4.3.2 of JCT 81 providing for written confirmation by the contractor of verbal instructions shall be deleted.

It follows that when the contract documents are being prepared that Clause 4.3.2 is lined through, or otherwise it will stay in and take contractual precedence over the employer's requirement entry, and that on signature both parties initial the clause deletion to confirm their acceptance.

This potential "grey" area between what is design development and what is a change in the employer's requirements can best be approached on the basis that the employer is entitled to know when he is being offered a new bargain by the contractor and he must have the option to buy or not to buy—or to negotiate. If the contractor fails to make it clear at the time when a new bargain is being offered he can hardly complain when his bill is subsequently challenged.

A further complication arises when the employer might perceive a functional re-arrangement, instigated by himself, as swings and roundabouts. This is particularly relevant in design and build projects simply because the contractual base-line is only conceptual, as opposed to the JCT 80 scenario of supposedly fully designed tender drawings and accurate bills of quantities prepared by the employer.

The contractor will on the other hand be much closer to the detail and will quite possibly be justified in claiming a change at additional *cost* to the employer, and with a possible *time* penalty.

If the design and build contractor plays fair and only calls "Change" for real functional or material selection issues which will put him to additional *cost* all well and good—the project should work out well for all concerned and the final account will be a non-contentious and relatively simple process.

Unfortunately one of the weaknesses of design and build is that there is great temptation for contractors, even more so than under JCT 80, to run with the hare and hunt with the hounds. Simply because the design and build contractor is in control of the design development process, and the employer through his agent is only an occasionally invited guest to the party, the temptation is for contractors to:

- Claim changes and justify same, deliberately moving or obscuring the original base-line; or
- Make "hidden savings" and not deal even-handedly with the employer in owning up to such savings with the same alacrity as when claiming changes and additional costs.

Thus the contract sum analysis becomes probably the most important of the contract documents—holding the key to both the identification and proper valuation of Clause 12 changes.

The employer and his agent are therefore only as powerful as the safeguards they write into the employer's requirements and whilst not generally condoning alterations to the well-considered and substantially proven JCT 81 Standard Form of Contract, I believe a few amendments are justified, with legal advice, to provide for effective post-contract monitoring of fair play.

Certainly I would strongly recommend the separate provision in the employer's requirements for the right of the employer to call for an audit, which in practice could be a mid-term and final account audit, even where public funds are not involved. Hopefully this will not be necessary in practice, but it is far from unknown for employer's agents to be found to be taking a passive role and allowing the contractor to make the running in deciding what "instructions" are proper changes as opposed to design development, whilst still reporting to their client, the employer, that somehow the books will be balanced on final account!

Some design and build contractors have well-established procedures for converting verbal instructions into Clause 12 changes, supported in some cases by the "game" of issuing Appendices to Minutes of Meetings after the event—and surprise, surprise, one of these Appendices, so often unread due to other pressures when stapled unannounced behind the formal issue of the Minutes, lists all perceived "Changes" together with budget prices. If not properly appraised and debated at the time, such "For Information Only" contractor's advice can soon assume contractual significance in its own right.

4.6 THE SUPPLEMENTARY PROVISIONS OPTION

The basic JCT 81 contract does not provide for pre-agreement of *cost* or *time* issues in relation to Clause 12 changes, but leaves such matters to be mutually resolved between the employer, through the employer's agent, and the contractor. Equally, it does not provide for any "first-fix" dispute procedure by way of third-party adjudication in the event of failure to agree.

As such the basic JCT 81 contract relies on trust and willingness to agree between the parties, hence it is a "soft" contract. This is in no way a disparaging term but merely a distinction between a contract which does not provide a strict change-order and disputes-control procedure, and a "hard" contract which does, i.e. where the contractor is required to take risks up front in taking on changes to the employer's requirements.

There may well be a premium price to pay for the employer, but at least he has the opportunity to reconsider his request for a Clause 12 change when told the price and time implication, if applicable.

This "hard" procedure is provided for by the JCT 81 Supplementary Provisions option, which is fully discussed in Chapter 5. If this optional procedure is adopted it provides a very useful discipline on employers in only considering necessary alterations, and it requires sensible handling by the contractor. It must not be used as a "pistol to the head" by way of quoting unreal prices or time consequences for requested changes to the employer's requirements.

4.7 TYPICAL PROBLEMS

Problem A

Employer's requirements, contract particulars section, Appendix entries on one project read:

"*Supplementary Provisions: To apply as amended below:*

  S1 Name of Adjudicator—To be agreed"

with no further provision.

The contract as signed, Appendix 1, read:

"*Supplement Provisions: to apply/not to apply*

  Name of Adjudicator: To be agreed."

The design and build contractor then ran an argument, when the final account became contentious, that:

1. Because *neither* of the "to apply/not to apply" options had been deleted in the Appendix of the signed contract that the Supplementary Provisions were *not* part of the contract, despite the clear intention to agree an adjudicator if necessary.

2. Alternatively, if he was wrong on his first contention, then as the S.6 pre-estimating procedures in respect of *cost* and *time* had *not* been enforced by the employer's agent during the first nine months of the project they had *de facto* been dispensed with.

3. Alternatively, if he was wrong on both of the above, then he had given budget costs each month by way of the Appendix to the Monthly Contractor's Report for all Confirmation of Verbal Instructions, so had in fact complied with the spirit of S6 in respect of *cost*, albeit no supporting details had been given, or advice on *time*.

The employer's agent and his in-house legal department, who had been responsible for drafting and supervising the signing of the contract documents, had therefore left the door ajar simply by not deleting either of the options in Appendix 1—"to apply/not to apply", an easy mistake and something which I suggest the JCT might like to prevent by providing that the Supplementary Provisions apply *unless* deleted by the parties.

Problem B

A multi-party development deal for a North London distribution centre where the building contract was to be JCT 81, but where there were four interested parties, namely:

1. Funder
2. Developer
3. Contractor
4. Secured tenant

The three known parties prior to tender were each named and represented by London solicitors, none of whom had a particular reputation as construction lawyers.

A quick pre-tender audit revealed the following problems:

(a) Some 75 mm thickness of tripartite amendments to the Standard Form of Building Contract With Contractor's Design 1981 Edition.

(b) No definition as to who would act as "employer's agent" in the contract particulars section of the employer's requirements.

(c) An old English form of bond, somewhere between a performance bond and an on-demand bond, enclosed as an Appendix, yet the contract particulars stated that a bond would *not* be required, in lieu of which the contractor would be required to give a parent company guarantee.

(d) Supplementary Provisions to apply with no provision for naming the adjudicator at S1.1. The fall-back procedure of S1.5 then kicked in and it fell to the arbitrator to appoint the adjudicator. Unfortunately the contract particulars then provided that Clause 39 arbitration would be deleted in favour of litigation, so at a stroke the mechanics for appointing the adjudicator disappeared.

Who said too many cooks spoil the broth? But at least we caught this one in time.

There is thus no substitution for understanding the basic principles and requirements of JCT 81. Anyone who makes other than strictly necessary amendments to the various Standard Forms of Contract, without specialist construction lawyer advice, does so at their own and their client's peril and potentially puts their own PI policy on the line.

The various Standard Forms of Contract are discussed later in this book. In my experience design and build is a very specialist area and only when a solicitor or counsel has had to litigate or arbitrate such a contract do all the subtleties begin to be understood. If, therefore, you have a design and build problem, do go to one of the known "names" in this field.

CHAPTER 5

THE SUPPLEMENTARY PROVISIONS

5.1 THE "HARD" CHOICE

For the very reasons outlined in previous chapters, there was a perceived dissatisfaction from employer organisations with JCT 81 as a "soft" form of procurement favouring contractors generally, particularly those who abused the system. Accordingly the British Property Federation made representation to the JCT drafting committee and a compromise was reached—the employer could choose a "soft" contract as per the basic JCT 81 provisions or he could opt for a "hard" contract by incorporating the JCT Supplementary Provisions S1 to S7, which for ease of reference are reproduced at Appendix A.

Whichever option is exercised it must be made clear from tender stage, and once the contract is awarded, if the Supplementary Provisions are to apply, they must be observed and enforced, if necessary. These Supplementary Provisions need to be read in detail and fully understood, but in synopsis form can be summarised and commented upon as follows.

5.2 ADJUDICATION (S1)

S1.1: A "first-fix" dispute resolution procedure up to practical completion, subject to the right of either party to have the dispute finally resolved by arbitration under Clause 39, or by litigation should Clause 39 be deleted, the adjudicator being named in Appendix 1, or to be appointed under S1.5.

S1.2: The definition of dispute is wide and "Adjudication Matters" cover all aspects of performance, *cost*, *time* and *quality*, except the tax deduction and VAT provisions of the contract.

Comment:

1. In respect of the nomination of the adjudicator the employer can:

   (a) Name an individual;
   (b) Leave blank: see S1.5;

(c) Name an institutional appointing body such as the Royal Institution of Chartered Surveyors;

2. However, as in the appointment of arbitrators, there is a school of thought which says the better the devil you know than the devil you don't know. Perhaps the happiest solution is none of the three above, but to insert "To be agreed by the Parties at time of contract award". In this way there is an almost certain chance that a well-respected individual, known to both parties, will be agreed at the outset.

3. If this action is overlooked in the excitement of getting the project under way the likelihood of mutual selection will diminish and S1.5 will apply if no individual is or can be agreed.

4. As all disputes relate directly or indirectly to money and as most disputes are likely to be contractor/subcontractor interim payment squabbles, arguably a quantity surveyor is the most appropriate discipline to appoint as adjudicator.

5. Whilst understanding the concept of a third-party individual to give a holding decision and thus enabling the works to proceed without too much loss of face, I question why an adjudicator should be *ex officio* immediately practical completion is granted.

6. It is often in matters of final acceptance, i.e. *quality* and compliance issues, as well as final account details, that disputes arise and naturally most of these occur post-practical completion.

7. I would therefore suggest that the employer's requirements should usefully extend the role of the adjudicator through to final completion and settlement of the final account as a total "first-fix" dispute resolution procedure.

S1.3: This sets the timescale for notification of a dispute and the action required of the adjudicator, following whose decision either party has 14 days in which to notify the other that the adjudicator's decision is not acceptable in the long term and thereby requires the issue to be referred to arbitration. In the meantime the adjudicator's decision is a provision of the contract and must, if possible, be followed.

S1.4: Any further dispute arising out of an adjudicator's decision is itself a dispute and referable to the adjudicator.

Comment:

1. Bearing in mind that most disputes post-contract will directly affect monies properly payable on interim valuation whether under Alternative A—Stage Payments, or Alternative B—Periodic Payments, time is of the essence in the adjudicator giving his "first-fix" decision.

2. Thus if Alternative B is being used and payments are monthly, the whole adjudication process needs to take less than a month from notification of dispute through to adjudicator's decision.

3. However, if the dispute is one of alleged non-payment by the main

contractor it will often take the subcontractor a week or two to realise he has not been paid *and* be able to confront the contractor.

4. It is therefore important that no time is lost in the appointment of an adjudicator when a dispute does arise, and accordingly this factor militates against nominating an institution as the adjudicator-appointing body at Supplementary Condition S1.

S1.5: A business clause providing for the eventuality of the death or other non-availability of the nominated adjudicator, or the passing on of the nomination if the original nominee feels unable to act.

Comment:

1. In the case of the original nominee's inability to drop other commitments at a moment's notice, or in the case of physical incapacity, other than death, it seems sensible that the original nominee renominates and that the two parties have no say in the matter.

2. In this event it should be made clear whether such a renomination is a temporary arrangement due to, e.g. illness, another case, holiday, etc., or whether the original adjudicator is in fact resigning as adjudicator under the contract.

3. If the latter scenario, it might be better if the original adjudicator writes to both parties, explaining his change of circumstances, such that the parties themselves have a chance of agreeing an alternative adjudicator, preferably before any dispute is notified.

4. Whether in the event of a dispute arising and the parties being agreed that their first choice adjudicator has not applied himself as they had wished, they can jointly dismiss him and replace him with another, the contract is silent. However, on the principle of mutual agreement, I see no objection to this course of action—having heard of it occurring only in the case of an unsatisfactory arbitrator.

S1.6: Given a dispute and reference to an adjudicator, each side shall be responsible for its own costs.

S1.7: Allows the adjudicator to secure his fees in equal proportions from each party before issuing his decision.

Comment:

1. On the face of it an eminently sensible provision.

2. However, it does not provide for the circumstance where one party is hostile to the whole procedure, e.g. a contractor unwilling to pay a subcontractor and who can then frustrate the adjudicator's decision by simply not paying his element of the fee.

3. A more practical provision might be to provide that the adjudicator's fees are to be paid by the party making the application, but that 50 per cent is a cost payable by the other party under the contract.

5.3   SUBMISSION OF DRAWINGS ETC. TO THE EMPLOYER (S2)

S2: Procedural points on the submission of drawings etc. to the employer, the most relevant being S2.2 where it is emphasised that the contractor's liabilities and obligations under the contract remain notwithstanding any comments or lack of comments by the employer, unless specifically stated.

Comment:

1. The short Sub-Clause S2.2 is not to be overlooked and puts the onus squarely on the contractor to state his case at the time of drawing or material approval by the employer or his agent for a Clause 12 change.

2. If the contractor takes the employer's comments and requests for fine tuning without demur he accepts a "no change" situation.

3. The important point is that under Clause 4.1.1 there is provision for the contractor making reasonable objection in writing to the employer in respect of any requested Clause 12 change, i.e. the contractor is deemed to be able to identify any difficulties, particularly programming implications.

4. Any employer's comments adopted by the contractor cannot therefore affect *cost* or *time* unless flagged up and agreed before the work is put in hand.

5.4   SITE MANAGER (S3)

S3: Merely stiffens the basic provision of Clause 10 in respect of there being a full-time resident site manager as opposed to a person in charge and calls for proper records to be kept as referred to in the employer's requirements and for the records to be available at all reasonable times to the employer or his agent.

Comment:

1. Notwithstanding the liabilities and obligations contractually placed on the design and build contractor, the employer must make proper provision for satisfying himself as to site procedures, e.g. labour records and photographs, quality control matters, ground conditions, concrete cube tests, fire certificates and progressive testing and commissioning.

5.5   PERSONS NAMED AS SUBCONTRACTORS IN EMPLOYER'S REQUIREMENTS (S4)

S4: Makes provision for the employer identifying specialist work and naming in the employer's requirements a preferred subcontractor, an option not provided for under the basic JCT 81. Equally the contractor may object and there is a renomination procedure as well as determination provisions.

Comment:

1. This is a useful and necessary provision if the employer is to control critical elements of the work, but the naming of the subcontractor must be done up front at time of tender.

2. Equally if third-party specialists are to be brought in direct by the employer during or overlapping the main contract, e.g. for information technology cabling, the employer will likely have to pay a premium both in *cost* and *time* if he does not use S4 to name his preferred specialists in the employer's requirements, and to put the contractor on notice pre-contract that such work by others must be accommodated.

3. It must also be remembered that if the contractor merely seeks the employer's consent under Clause 18 to subcontract work, the employer will normally have no direct access to that domestic subcontractor, but in the cases of "named subcontractors" there are inevitably informal links from pre-contract discussions.

5.6 BILLS OF QUANTITIES (S5)

S5: Provides for the employer preparing a bill of quantities as part of the employer's requirements, provided the method of measurement is stated. It follows that this will then become the base document for interim valuation and final account purposes.

Comment:

1. A particularly unhelpful clause in my opinion, as it appears to sanction employers preparing bills of quantities in the usual way as under JCT 80 and then passing the problematic design baby over to whichever tenderer is prepared to adopt it for the lowest price, i.e. "risk dumping".

2. The question then arises: What if the contractor then comes up with a different design solution or a different method of construction?

3. Unless a revised bill of quantities can be remeasured and agreed prior to work progressing too far, so creating a new base-line, it is very difficult to see how the strict change procedure of the Supplementary Provisions S6 and S7 can be operated.

5.7 VALUATION OF CHANGE INSTRUCTIONS (S6)

Supplementary Provision S6 is *the* centrepiece of the Supplementary Provisions and requires detailed analysis:

S6.1: Specifically overwrites the basic JCT 81 Clause 12, Changes, Clause 25, Extension of Time, and Clause 26, Loss and Expense, and substitutes the following:

S6.2: Provides for the contractor submitting, within 14 days of the date of the relevant instruction, estimates of *cost* relevant to Clause 12, *time* relevant to Clause 25 and direct loss and/or expense relevant to Clause 26 unless:

- Such period other than 14 days has been agreed, or in the event of disagreement whatever period may be reasonable in all the circumstances;
- The employer states in writing at the time of issue of the instruction, or within 14 days thereafter, that estimates are not required;
- The contractor on his own behalf, or on behalf of a subcontractor, raises objection within 10 days to the effect that the provision of all or any such estimates is unreasonable.

Comment:

1. S6.2 does not require the contractor's estimates to be in writing and I believe this to be an oversight on the part of the JCT drafting committee.

2. S6.2 is of course primarily designed to allow the contractor to alert the employer to the fact that his instruction has a cost and time implication.

3. However, the full text of S6.2, second line, also provides for the employer having the right to issue an instruction and within 14 days to call for estimates by the contractor, but this can work both ways, i.e. the contractor might see the employer's instruction as "swings and roundabouts" or even a hidden saving and keeps quiet, but the employer can call for an estimate to flush out his perceived saving.

4. If they then disagree that there is an addition or a saving in *cost* or *time* or that there are knock-on effects prejudicial to the contractor, either party can call in the adjudicator (S1.1).

5. As regards the timing provision and what happens if they cannot agree the period for provision of estimates, I believe the S6.2 wording is too woolly—I would suggest striking out the third option on the basis that if they cannot agree then the period should be 14 days.

6. What then if the contractor or a subcontractor takes exception to providing estimates? How does the employer's agent really know if the objection is valid and if *he* does not know, being close to the project, how can an adjudicator be expected to have better knowledge?

7. In practice the answer is that neither party usually wants a third party to appear on the project and potentially tell them they are wrong, so usually the mere threat by one party to call in the adjudicator refocuses the mind of the other party, and a compromise is reached.

8. Again S6.2.2 fails to require the contractor's objection to providing estimates to be in writing—this is I believe another oversight and should be amended.

S6.3 Lays down the ground rules for the contents of the Contractor's estimates in respect of *cost, time* and claims.

Cost

S6.3.1: "the value of the adjustment to the Contract Sum, supported by all necessary calculations by reference to the values in the Contract Sum Analysis;"

Time

S6.3.2: "the additional resources (if any) required to comply with the instructions."

S6.3.3: "a method statement for compliance with instructions;"

S6.3.4: "the length of any extension of time required and the resultant change in the Completion Date;"

Loss and expense

S6.3.5: "the amount of any loss and/or expense, not included in any other estimate, which results from the regular progress of the Works or any part thereof being materially affected by compliance with the instructions under Clause 12."

Comment:

1. These ground rules are quite rightly definitive but in practice they do depend upon firstly the detail available in the contract sum analysis documentation *and*, secondly, on common sense between the parties.

2. If the parties can agree Clause 12 changes without going into the full detail required by S6.3 so be it, but they may have to account to an auditor at some later date, so there is no safe substitute for full compliance with the S6.3 Sub-Clauses, however brief the back-up notes might be.

S6.4: Assumes agreement will be possible within 10 days of the receipt of the contractor's estimates in which case such agreement will be binding.

Comment:

1. A good example of the need to define a "day". What if the contractor delivers such estimates late on a Friday? Ten *calendar* days is close of play Monday week, i.e. six *working* days, assuming the Monday is not a Bank Holiday.

S6.5: Provides for the consequences of any failure to agree the contractor's estimates, whether in part or in whole, and is robust in that it gives the employer control of the situation. He has three options:

S6.5.1: He may instruct the work to proceed but leaving *cost, time* and any loss and/or expense claim to be agreed after the event; or

S6.5.2: He may withdraw the instruction, but must pay any reasonable design costs incurred by the contractor in responding to the specific instruction now withdrawn; or

S6.5.3: He may refer the matter to adjudication under S1.

Comment:

1. It should be noted that in the event of failure to agree the cost of Clause 12 changes only the employer may call for adjudication—if the contractor considers he is being put upon he can call upon Clause 4.1.1 and his rights of reasonable objection, which is an "Adjudication Matter" as defined in S1.2.5, but he cannot refuse to carry out the instructed work simply because he and the employer cannot agree *cost*.

2. As under Clause 4.3.1 all instructions issued by the employer shall be in writing, it is to be presumed that whichever option is exercised by the employer should also be in writing.

S6.6: Provides for the eventuality of the contractor not providing the estimates required under S6.2. This could be in respect of timing or sufficiency of detail and in the second sentence of the clause there is a very real sanction against the non-compliant contractor:

- No payment of any non-agreed Clause 12 changes until the issue of the final certificate;
- Contractor not entitled to recover the cost of finance for such non-agreed Clause 12 changes.

Comment:

1. This scenario assumes the contractor has put the work in hand without either providing the required estimates or having raised objection to the provision of such estimates.

2. If the contractor has made a submission, albeit incomplete, not sufficiently particularised or simply unacceptable to the employer, he will still be caught by this provision if he proceeds with the work *unless* he first obtains the employer's written instruction issued specifically under S6.5.1, i.e. the pre-costing requirement is dropped by the employer.

3. As the final certificate is not issued until the employer is satisfied as to the making good of defects and until all matters concerning the final account including any loss and/or expense claim have been settled the contractor could have to finance any unagreed but valid Clause 12 changes for many months, if not years.

5.8 DIRECT LOSS AND/OR EXPENSE (S7)

S7.1: Merely re-affirms that S7 provisions take precedence over Clause 26.

S7.2: Provides that the contractor, whether or not he has complied with the strict provision of S6.3.5 in providing a forward estimate of potential loss and/or expense after receiving an employer's instruction under Clause 12, must also provide an estimate of actual loss suffered at each interim payment, whether under Alternative A—Stage Payments or Alternative B—Periodic Payments.

Comment:

1. On the face of it, S7 would appear to cut across the strict provision of S6 and the requirement for the contractor to provide forward estimates of loss and/or expense.

2. However, JCT Practice Note CD/1B does make the point in relation to S6 that:

"The tribunal have always assumed that on all contracts placed on its Standard Form With Contractor's Design the Employer will give careful consideration in setting down his requirements whether this be in a broad or in detailed form and that during the course of the contract the number of Change instructions will be somewhat limited and far less numerous than the number of Variation instructions that often arise under its 1980 Standard Form of Contract.

Clause S6 has been prepared on this assumption and it is not intended that it be applied to all the Change instructions on a contract where there are a large number of minor Change instructions."

3. As such, S7 covers the fall-back position of where, on a collective basis, the contractor wishes to state his case for reimbursement of perceived loss and/or expense.

4. In practice the sooner any such notification is given the better, but in real life it is only when subcontractors submit claims to the contractor that alleged additional costs come through, and this is often months after the "List of Matters" defined under Clause 26.2.

5. Reasonably ascertaining facts as opposed to fiction then becomes a major task and much will depend upon whether the contractor and subcontractor are prepared to go "open book". As a rule, if they resist an "open book" approach they have something to hide and nothing can be taken on trust.

S7.3: Merely provides for the ongoing submission of further estimates by the contractor of loss and/or expense allegedly incurred, whether or not the employer has accepted the principle or the amount previously claimed.

S7.4: Provides a 21-day period for the employer accepting or rejecting the contractor's estimates under S7.2 or S7.3, with the following options:

S7.4.1 He accepts the estimate; or

S7.4.2 He wishes to negotiate the additional cost, and failing agreement he will either:

(a) Refer the matter to adjudication under S1.1, or

(b) Rely on the provisions of Clause 26; or

S7.4.3 He will dispense with the regular estimates and immediate adjudication option but let matters take their course under Clause 26.

Comment:

1. Again it is the employer who decides which option will apply, and the contractor cannot apparently call in the adjudicator himself.

2. If however Clause 26 applies and there is still no agreement of the additional costs claimed by the contractor, then the contractor can call for

adjudication, but only if the project has yet to reach practical completion, otherwise the adjudicator is *ex officio* under S1.1.

3. By default, if the employer fails to exercise any of the three options within 21 days he lets in the contractor's claim, so it is important for both sides to be clear whether they are working to calendar or actual working days, Monday to Friday excluding recognised holidays.

S7.5: Provides for adjustment of the contract sum upon the agreement or the adjudicator's decision of any matter arising under the Supplementary Provisions S7.4.1 or S7.4.2.

Comment

1. This only provides for adjustment of the contract sum in respect of loss and/or expense not previously agreed under S6, whereas Clause S6.3.1 provides for the contract sum adjustment in the event of pre-agreed loss and/or expense.

2. However, as a matter of housekeeping the contract sum also requires adjustment in respect of adjudication matters which are not referred to arbitration under S1.3.4, and therefore become binding, but no express provision is made other than S1.3.3 where "such Adjudicated Provisions shall be final and binding on the parties unless referred to Arbitration . . . "

S7.6: Repeats the provision of S6.6 in that if the contractor fails to comply with the submission procedure of S7.2 he takes his chances under Clause 26, but has in any event to fund such loss and/or expense until the final certificate.

Comment:

1. This provision again is intended to penalise the contractor who makes retrospective claims at the end of the project and should equally be used by contractors in imposing proper discipline on their subcontractors.

2. Alternatively it could be argued that it provokes a claims culture, but at least it should force the contractor to table any justification at the proper time and give the employer's agent a more realistic opportunity to deal with matters on their true merits.

5.9 THE TWIN GOALPOSTS: COST AND TIME

The ideal to be achieved throughout the project is certainty of both *cost* and *time* which should benefit the employer and contractor alike. These twin goalposts are first established by the contract terms, i.e. the contract sum and the agreed start and completion dates.

If, by reason of Clause 12 changes, there is any question of moving either of the twin goalposts then the Supplementary Provisions provide the mechanics for so doing.

Any employer's agent who allows the contractor to plead that the pre-estimate procedure will delay the project and relaxes the S.6 requirements starts down the slippery slope of loss of control.

However, with the best of intentions the operation of provisions S.6 and S.7 can result in genuine disagreement between employer and contractor as to the proper *cost* and *time* implications of Clause 12 changes. In this event, it is wise to use the Adjudication provision at S.1. History suggests that rarely does the losing party in an adjudication risk getting the same result in arbitration and having to pay out real money, plus the other side's costs, in so doing.

Summarising, therefore, the JCT 1981 Supplementary Provisions provide the option of certainty for two of these essential elements previously identified as critical to any form of contract. If *cost* and *time* can be pre-agreed and amended as necessary on a rolling basis without major disagreement, all parties in the contractual chain, from developer and funder down to sub-contractor and sub-subcontractor must benefit.

The JCT Supplementary Provisions should be a requirement for any design and build project, in my view. In fact, it could be arguable that *not* to include the Provisions amounts to a breach of duty, particularly where a solicitor or other professional consultant is advising a lay employer.

Minimisation of the risk of disputes must be a common objective, and as such of the three elements only *quality* is elusive.

CHAPTER 6

PAYMENT PROVISIONS

6.1 THE CONTRACT SUM ANALYSIS

Essential functions

From the employer's point of view the contract sum analysis is like having your birthday and Christmas in one—what you do not ask for when you are being treated you will have to buy for yourself later in the year. From the contractor's point of view the contract sum analysis is a real-life double whammy—which if not handled properly at the outset will come to haunt the contractor all the way through the project up to the final account and possibly beyond.

As such, under the JCT 81 the contract sum analysis is a pivotal document whose primary functions are:

- Confirmation of contractor's bid in terms of scope and *quality*;
- The vehicle for agreeing the value of any changes in the employer's requirements.

The secondary functions of the contract sum analysis under JCT 81 are:

- The calculation of interim valuations—whether Alternative A Stage Payments or Alternative B Periodic Payments apply;
- The calculation of fluctuations in price of labour and materials under the formula rules where Clause 38 applies.

Interestingly JCT Practice Notes CD/1A and 1B cover only three of the four functions identified above, omitting to mention the very first, i.e. the confirmation of the contractor's bid in terms of scope and *quality*.

Judging tenders

To explain this important point let us assume we have sent out employer's requirements calling for the return, by way of competitive tenders, of the following submissions by all tenderers:

- Contractor's proposals, including developed drawings;

- Form of tender, including undertakings as to required forms of bond and warranties;
- Contract sum analysis, in a required elemental form and left blank for pricing by each tenderer.

Without the latter, i.e. a precise statement by the employer of the form of contract analysis required as a condition of tender, setting out the level of detail to be submitted in support of the tender, the employer and his team have very little other than bottom-line price as a basis of tender evaluation.

On JCT 80 there is a developed design, measured and stated by the employer, in the form of bills of quantities and priced by the contractor as part of the tender return. On JCT 81 there is usually no common developed design and therefore no bills of quantities prepared by the employer.

Accordingly design and build tenders are essentially a combination of:

- A design and buildability competition;
- A total cost bid, i.e. build cost plus design fees,

and where *time* is usually not a problem as a tender variable, but the big unknown is *quality*.

How does the employer begin to answer this last and important question? What if the second or third contractors in terms of *cost* are going to provide a better scheme and a more comfortable or sophisticated or glamorous or efficient building than the lowest tenderer?

I have deliberately used some typical loose and subjective qualities to make the point that each client, and potentially each individual in the employer's tender appraisal team, has imprecise objectives and understanding of what is to be achieved in the finished product, yet is now being asked to make the two most important decisions on any project:

- Can we now commit to proceeding with the project?
- If so, which contractor is likely to give us the best value for money?

"Value for money" is a subject in its own right, and as identified in the Latham Report, in the past insufficient attention has been paid by employers in commissioning buildings as to the potential *cost* penalty of accepting the lowest bid. *Quality* and in particular *costs* in use versus maintenance budgets are relevant considerations in any tender situation, and none more so than when public funds are involved.

So we come back to the typical design and build tender conundrum: how can the employer and his team know whether one tender offers more or less in *quality* terms than another? Indeed the same question can be extended to the scope of the work. The contractor's proposals may well state various inclusions, qualifications or positive exclusions, but with hindsight there are usually "grey" areas.

Tender ground rules

The answer is for the employer to lay down the ground rules in the tender enquiry by stating in the employer's requirements that:

Rule 1:

- All tenders must be accompanied by a fully priced contract sum analysis, as issued in blank form, which totals to the tender sum. (A typical contract sum analysis proforma is reproduced at Appendix B.)

Rule 2:

- That as part of the tender evaluation process the employer reserves the right to call for the submission within two working days of a fully detailed typed breakdown to support all lump sums showing in the contractor's contract sum analysis.

The above procedure is in reality no different from the JCT 80 procedure of calling in the priced bills of quantities from the two lowest tenderers, except that in the absence of bills of quantities prepared by the employer and issued in blank form as a tender document, under JCT 81 it is left to the tenderers to quantify their own design solution and be prepared, on demand, to submit mini-bills or builder's quantities as they alone see fit in support of the obligatory contract sum analysis.

Tender analysis: first stage

Once he has received the priced contract sum analysis under Rule 1 the employer should be able to look behind the bottom-line price, and the developed drawings as submitted, to examine the scope of work and the *quality* offered by those tenders which appear on price alone to be the most favourable. This is done by applying quantity surveying cost-planning techniques, but in reverse.

On design and build projects where the building footprint is stated as part of the employer's requirements the gross internal floor area (GFA) is common to all tenders. The GFA is by accepted definition the measurement on plan of the area within the internal face of all external walls, measured over internal walls and partitions.

A tender analysis proforma should therefore be set up by the employer as the initial comparison document, listing each building element and sub-element of the required contract sum analysis down the left hand side. Across the top in four-sub-column format are listed however many tenders as are worthy of initial consideration. (A typical employer's tender appraisal is reproduced at Appendix C.)

In the first column under each tenderer is the building element value copied from the already returned contract sum analysis, totalling the tendered sum, and in the second column is the elemental price per square metre, taking the common GFA as the divisor. In the third column the building sub-element values, as copied from the contract sum analysis and totalling the tendered sum, are stated. Then in the fourth column, the sub-elemental price per square metre is stated—again using the common GFA as the divisor.

On projects where the design brief is more flexible and there is no stated building footprint as part of the employer's requirements the same cost planning in reverse principle still applies. Measure the GFA from each tenderer's submitted scheme drawings and apply that in each case to the total declared elemental and sub-elemental values, such that comparative prices per square metre are established.

In this way it should readily be apparent as to:

- Whether all tenderers have, subject to the vagaries of their submitted scheme, included for all required work, i.e. scope check;
- How each tender compares by way of price per square metre at elemental level: this should be within acceptable cost bands as may be found in recognised cost books, e.g. Spons, Laxtons, Griffiths, Wessex or the Building Cost Index Services (BCIS);
- How each tender then compares at the next level down, i.e. subelement level by way of price per square metre, to be compared with industry norms as above.

Inevitably attention at this initial stage will be focused on the two lowest tenders. The comparative results obtained by analysis in the format detailed above should then be compared with the notional tender represented by the employer's elemental cost plan.

Once the above process has been completed and all arithmetic crosschecked it will be possible to draw up a check-list for each tender identifying:

- Gaps—potential mistakes and misunderstanding of the employer's requirements;
- Misunderstandings in respect of the employer's requirements by way of spotting "rogue", i.e. high or low, elemental or sub-elemental unit rates.

However, an ascertained unit rate can still be misleading in that any unit rate is necessarily a combination of *quantity* and *quality*.

At this initial stage no declared quantities are on the table other than the measured GFA, so any tender analysis and comparison with the employer's cost plan cannot be definitive, i.e. *quality* can still be at large—as indeed can be the scope of work in that the contractor's perceived quantities, which are the first factor in the pricing equation, may or may not relate to the drawn information supplied by that contractor.

Tender analysis: second stage

It is therefore essential that there is a second stage to the employer's tender analysis, focusing on no more than three tenders if reasonably close, but often involving just the lowest tender: hence the need for Rule 2.

It is equally essential that this second stage appraisal is provided for as part of the "rules of the game", i.e. as a required provision in the employer's requirements document, as mentioned previously. The lowest tenderer should not therefore be caught by surprise and put in a false position, as has happened on occasions under JCT 80 where the lowest tenderer, or second lowest tenderer have had their bills of quantities called in, but for whatever reason the bills have not been priced, so have had to be created post-haste!

Assuming the back-up documentation to the contract sum analysis on a design and build project under JCT 81 has been called for under Rule 2, and is then produced, it will usually take the form of builder's quantities, with different trades quite possibly showing different levels of detailing, having been subcontracted out.

There will be no stated "Method of Measurement" unless the employer as part of the tender ground rules has prescribed a short form standard method such as the RICS Method of Measurement for International Works.

However, whatever else the back-up documentation shows, or does not show, it does need to show separately:

- Preliminary costs, in reasonable detail such that "fixed" costs, i.e. site set-up and site clearance costs can be distinguished from monthly "running" costs;
- Design costs, split between
  (a) Pre-contract
  (b) Post-contract design development including provisional sum expenditure as may be instructed
  (c) The percentage required to be added to any omitted or additional work, i.e. Clause 12 changes;
- Overheads and profit—how costed;
- Provisional sums, preferably included and totalled within the relevant elements;
- Dayworks: the stated provisional sum values as given in the employer's requirements, with the required definition, and extended by the contractor's required percentage mark-ups in respect of labour, materials and plant, again with a separate (and hopefully lower) percentage addition for design fees, as opposed to pre-planned Clause 12 changes.

Back-up documentation

So we come to the design and build contractor's real-life double whammy referred to in the first paragraph of this chapter: Tender Rule 2.

To support his tender and the official contract sum analysis, which is a defined contract document under JCT 81, the contractor now has to show a detailed back-up for each sub-element.

The big dilemma is: in what detail does the contractor declare his contract sum analysis back-up? And what will satisfy the employer and the employer's quantity surveyor?

Ideally the design and build contractor, with the help of his architect, will be able to reasonably show a detailed multi-item build-up for each sub-element value, i.e.:

- Quantity of work in each declared item;
- Brief description of the item, including some hint of *quality*;
- Unit rate.

If so, all well and good, but what if the contractor has only partly developed the employer's stated concept design and has had to go out to tender for the mechanical and electrical packages (often 40 per cent of the total price) on a cover-price basis? How can he do any better than single-line lump sum values for each sub-element, e.g. hot and cold water, heating, air conditioning, etc?

Provided the contractor's proposals document either confirms the mechanical and electrical systems required in the employer's requirements, or is specific as to the systems being offered against the employer's performance specification, the all-important *quality* definition is spoken for—but conversely, if not, and the employer accepts the contractor's tender, he takes an awful lot on trust and chance.

But continuing the mechanical and electrical cover-price scenario, the design and build contractor is still not out of the woods even if the contractor's proposals tie down the systems to be delivered. What if the employer subsequently requires significant alterations in room layouts, but the only contractual definition of value is, say, six single-line entries in the contract sum analysis?

The other half of the contractor's double whammy is that he only gets one bite at the cherry and he therefore has a dilemma at this critical second appraisal stage. Does he go in on a minimal detail basis, knowing that if Clause 12 changes are subsequently instructed he has limited scope for cost recovery with the employer, but with a significant risk of being non-back-to-back with his yet-to-be-selected subcontractor? Or does he guestimate the likely quantities and do a balanced work-back such that, warts and all, his single-line sub-elemental values are apparently justified by quantities and unit rates?

If he opts for the second method, at least he has a better chance of being able to justify recosting of any changes as may be instructed under Clause 12, as no employer or employer's agent will happily accept a design and build contractor's claim to alleged additional costs unless the contractor can reasonably show:

- What was in the original scheme; and
- How that was costed in relation to the declared elemental or sub-elemental value stated in the contract sum analysis.

Evaluation of changes

When one is faced with Clause 12 changes and there is no reasonably detailed back-up to the contract sum analysis there is inevitably no definitive drawn information to show what was required at time of tender. The contractor may well have submitted a fully detailed drawing as to what is supposedly now required in order to comply with the alleged, or admitted, change in the employer's requirements, supported by a remeasured sub-bill more akin to a full measurement under Standard Method of Measurement, Edition 7.

Whilst the latter may be applicable under JCT 80, it is certainly inapplicable under JCT 81, i.e. there must be parity in quality of information and basis of costing as between any design and build contractor's Clause 12 change submission and the back-up detail supporting the contract sum analysis.

A good example of this point, as well as an example of the risks a design and build contractor takes in the level of detail described in his contract sum analysis, would be in respect of mechanical and electrical services. Every project will require builder's work items, i.e. holes through walls for pipes, chases, etc. and most projects will require fire stopping between fire compartmentation elements, e.g. services riser ducts through floors and openings through fire walls.

On the overriding principle that the design and build contractor has allowed, somewhere in the detailing of his pricing that goes to make up the agreed contract sum, for elements of work which are required to comply with the Building Regulations or other express terms of the specification as incorporated in the employer's requirements, the question is: where is that "somewhere" if not stated in the back-up information supplied in support of the contract sum analysis?

The answer can only be that it is included in the general level of pricing, i.e. in the unit rates disclosed by the contractor. It follows that if a Clause 12 change is instructed requiring more holes through walls because there is more pipework and there are no items in the contract sum analysis back-up document for holes or for fire stopping, the contractor is deemed to be reimbursed by the quantity adjustment applied to the unit rate for the pipes as used in the tender.

What is inadmissible is a five-page remeasurement of all holes and fire stopping, i.e. lower level detail not stated in the contract sum analysis and the back-up thereto, but nevertheless implicit in the original scheme.

Where there is then no attempt to give credit for what must always have been included in the original scheme, the contractor is guilty of duplicity and a flagrant abuse of the principles, if not the written rules, of design and build.

Contractual status

So we return to the opening point—an employer must be clear as to the purposes of the contract sum analysis as required by the JCT 81 contract. He should understand that *by itself*, i.e. usually an elemental and sub-elemental summary as required for the first phase tender return, giving lump sum values only, it will fail on all four counts in that:

- It will not allow proper examination of tenders as regards scope and *quality*;
- It will require substantial guesswork using ball-park percentages at each interim valuation—assuming Alternative B is being used;
- It will be virtually impossible to satisfactorily agree any Clause 12 changes as instructed, including the expenditure of provisional sums;
- It will not easily allow the formula adjustment of fluctuations under Clause 38, if required, unless the blank contract sum analysis is formatted into the necessary index categories.

It therefore bears repeating that according to the definitions of JCT 81 only the contract sum analysis is a contract document. Unless the parties agree the back-up document, which for ease of reference can be called the mini-bill, and specifically incorporate it in the definition of the contract sum analysis at Article 4 of Appendix 3, it will be arguable as to what reliance can be placed on the mini-bill.

Mini-bill or schedule of rates

As to whether the contract sum analysis back-up document should be accorded the status of a mini-bill or simply a schedule of rates, this will depend in the first instance on whether the contractor has volunteered quantities, or failing that, whether the employer has insisted on quantities being disclosed prior to acceptance of the tender.

If no quantities have been disclosed—and the tender contains, say, just items and unit rates—it is unquestionably a schedule of rates, but unless quantities are disclosed and each item mathematically extended, how can the employer know whether it all adds up to the tendered sum, which if agreed will become the contract sum?

Equally, if the back-up document discloses no quantities, factor (A), but short descriptions for each item, with unit rates, factor (B), and extended values, factor (C), which collectively total to the lump sum declared in the contract sum analysis, it is only then a question of dividing the extended total (C) by the unit rate (B) to establish the missing quantity of item (A), when needing to assess value for interim payment or Clause 12 change purposes. So why not show quantities and save the convoluted arithmetic?

Accordingly the contract sum analysis back-up document needs to be, or to include, a mini-bill, showing quantities, with items described in sufficient detail as to reasonably establish *quality* and to enable the employer to verify the unit rates proposed. On one JCT 81 project the design and build contractor insisted on referring to such a detailed mini-bill as a "Schedule of Rates" and steadfastly opposed the employer's agent in placing any reliance on the disclosed quantities. When the counter-argument was put that if:

1. The contract sum as agreed was factually and mathematically the total of all the disclosed elemental and sub-elemental values in the contract sum analysis, and
2. The total value of each sub-element was factually and mathematically the total value of each item in the relevant section of the mini-bill,

then it was argued that:

3. On a line-by-line, item-by-item basis, if value (C), the total value, was contractually fixed and value (B), the unit rate, was contractual, then as a matter of logic value (A), the quantity, must also be contractual for final account purposes.

That particular design and build contractor insisted on spending good money with solicitors. Even our counsel managed to advise that the contractor's argument was not entirely without merit and wrote a multi-page advice note —all of which goes to emphasise the need for clarity of purpose when setting up a design and build contract. In other words, how will interim valuations and Clause 12 changes be assessed?

Such are the trials and tribulations of design and build: JCT 81 lends itself naturally, even more so than JCT 80, to disputes if the parties are so minded.

The key to any successful design and build contract is therefore two-fold:

- Set up the employer's requirements correctly in the first place, particularly the contract sum analysis and supporting documentation.
- Get to know the contractor at employer's agent level, building a working relationship on the basis of mutual trust as soon as possible, even if it means having an early falling-out to clear the air.

Typical forms

It remains therefore, by way of summarising this important section, to again refer the reader to Appendices B and C. At Appendix B is set out a typical blank form of contract sum analysis as one might expect to see sent out as part of the employer's design and build tender enquiry. This should then be filled in by each tenderer and returned in support of the total entered on the Form of Tender. At Appendix C is set out a typical employer's elemental tender comparison, previously referred to as cost planning in reverse.

Whereas each project may require its own form of contract sum analysis, good practice dictates that there should be a generally accepted standard proforma covering all possible projects. Accordingly, reproduced as Appendix D is the "Recommended BCIS (Building Cost Information Service) Elements for Design and Build", as published by the RICS.

At least if one opts to use this standard proforma one can be sure of comparing apples with apples, when considering elemental unit costs as tendered with current rates as published by the BCIS.

6.2   INTERIM PAYMENTS

No "third-party umpire"

It is in respect of matters of payment that the difference in philosophy between JCT 1980 and JCT 1981 With Contractor's Design usually manifests itself in real terms post-contract.

In the former procurement system, supposedly fully designed by the employer and his team, the architect is required to assume the role of independent referee or umpire as between employer and contractor, despite the fact that the architect is appointed and paid by the employer. This arrangement may seem like "Chinese walls" but as explained in Chapter 3 under "The Employer's Agent" this is a traditional arrangement which has over the years worked tolerably well.

However, with design and build as a procurement system the key role is taken by the employer's agent to perform the function of contract administrator and quite specifically in contractual terms he sits in the employer's camp—all as explained by JCT Practice Note CD/IB, Article 3, Commentary, and as discussed in detail in Chapter 3. There is therefore no "third party umpire" to independently certify interim payments between the parties.

Alternative methods of interim payment

Under JCT 1981 the interim payment procedure set out in Clause 30.1 is simpler than would appear at first sight and can be summarised thus:

30.1.1   Provides for either:

Alternative A: Stage payments, or
Alternative B: Periodic payments

30.1.2   Whichever alternative has been selected, the interim payment shall be the gross valuation, less:

- Retention
- Previous interim payments

Alternative A: Stage payments

30.2.A The gross valuation shall be made up of the following amounts, all subject to retention:

- Cumulative value when each stage is reached;
- Changes (Cl. 12.2) and provisional sum expenditure (Cl. 12.3);
- Fluctuations (Cl. 38).

30.2A.2 The gross valuation shall also include the following amounts *not* subject to retention:

- Materials (Cl. 8.3), royalties (Cl. 9.2) and the employer's right to set-off for defects (Cl. 16.2 and 16.3);
- Loss and expense (Cl. 26) and antiques (Cl. 34.3);
- Contributions, levy and tax, fluctuations (Cl. 36) and labour and material cost and tax fluctuations (Cl. 37) as payable to the contractor;
- Contributions etc. as above, but allowable to the contractor;
- Where Clause 38 fluctuations by formulae adjustment has been chosen, the costs of lift components, structural steelwork and catering equipment delivered, other than prematurely, to or adjacent to the works.

Alternative B: Periodic payments

30.2.B.1 The gross valuation shall be made up of the following amounts, all subject to retention:

- Work properly executed, any design work carried out, changes (Cl. 12.2) and provisional sum expenditure (Cl. 12.3);
- Materials and goods delivered, other than prematurely, to or adjacent to the works;
- Materials or goods which the employer directs shall be paid for prior to delivery to site;
- Fluctuations (Cl. 38).

30.2B.2 The gross valuation shall also include the following amounts *not* subject to retention:

- Materials (Cl. 8.3), royalties (Cl. 9.2) and the employer's right to set-off for defects (Cl.16.2 and 16.3);
- Loss and expense (Cl. 26) and antiques;
- Contributions, levy and tax fluctuations (Cl. 36) and labour and material cost and tax fluctuations (Cl. 37) as payable to the contractor;
- Employer's right to set-off for defects (Cl. 16.2 and 16.3) or as 30.2B.2.3 above, but allowable by the contractor to the employer.

The above summary of the alternative methods of interim valuation describes the work that qualifies for valuation at each stage or each month and states whether each element of the valuation is subject to the retention percentage as provided for in Appendix 1 at Clause 30.4.1. It does not deal with the procedure for originating the interim valuation and what happens if the prescribed procedure is not followed. Such matters are discussed later in this chapter.

Quality of work

Where serious problems can arise is in respect of the value properly payable and the provision under Clause 16.2 (Defects notified *at the end* of the defects liability period) and 16.3 (Defects notified *during* the defects liability period). In these circumstances, the employer's agent has granted practical completion. Commonly, practical completion is conditioned on certain snagging items or even more serious defects being made good by the contractor, at the employer's convenience, during the defects liability period, commonly 12 months. If such remedial work has to be done at the weekend or during the night, so be it, there is no claim on the employer for additional costs.

However, Clauses 16.2 and 16.3 give the employer the option of foregoing remedial works, but living with the notified defect. To exercise this option the employer must instruct the contractor (preferably in writing) that he is exercising this option and state the "appropriate reduction" he requires from the interim payment otherwise due.

If the option is exercised under Clause 16.2, say on re-inspection after 12 months, it may well be that all other matters, i.e. final account and claims may already have been resolved, and a further interim valuation may have been issued. Accordingly only the final moiety of retention will be due for release on a final certificate, so there may well be insufficient funds still held by the employer.

On the other hand, if the option is exercised under Clause 16.3, i.e. during the defects liability period, the employer will be retaining one half of the retention fund, *plus* the value of any further yet to be agreed changes and claims. It follows that the sooner the employer exercises the option the more secure his position is likely to be.

As to the position concerning perceived defects *before* practical completion, JCT 81 makes no express provision other than at Clause 30.2B.1 under Alternative B: Periodic Payments and where it provides that only "work properly executed" shall qualify for inclusion in the gross valuation calculation.

As in other forms of contract, the interpretation of whether work has been "properly executed" is a matter of judgement by the individual charged with the function of payment certifier, usually a quantity surveyor in the first instance, but formally the architect, engineer, contract administrator, or, in the case of JCT 81, the employer's agent. However, the big difference with JCT 81 is that there is no independent referee/umpire to blow the whistle or put the finger up, so to speak.

It is unfortunately a feature of JCT 81 that the less scrupulous contractors will try to substitute similar (supposedly) materials for those which they have proposed and have had approved by the employer's agent, or will offer a dubious standard of workmanship in the absence of a fully detailed material and workmanship specification dictated by the employer as in JCT 1980. Accordingly the employer's agent must be constantly alert that the standards are not being compromised, and the best time to pull the design and build contractor up is at interim valuation, should it be necessary.

The important function of certifier of both payment and *quality* is expressly given by the JCT 81 contract to the employer—not even to the employer's agent (see Chapter 3), although the function is inevitably delegated. There is therefore immediate potential for direct confrontation between employer and contractor if issues of *quality* and payment cannot be factually resolved on their merits.

Interim valuation procedure

So we come to the all-important interim valuation application procedure, which is a significant departure from JCT 80.

There being no named third-party certifier under JCT 81, the procedure is as follows:

"30.1.1   Interim Payments shall be made by the Employer to the Contractor in accordance with Clauses 30.1 to 30.4 and whichever of the Alternatives A or B (Stage or Periodic) i.e. Appendix 2 applies to this Contract.
. . .

30.3.1   The Contractor shall make Application for Interim Payment as follows:

30.3.1.1   Where Alternative A applies, Application for Interim Payment shall be made:
   • on completion of each stage set out in Alternative A in Appendix 2, and . . .
   • after the expiration of the Defects Liability Period named in Appendix 1, or . . .
   • on the issue of the Notice of Completion of Making Good Defects (whichever is the later).

30.3.1.2   Where Alternative B applies, Application for Interim Payment shall be made:
   • at the Period for Applications for Interim Payment stated in Alternative B in Appendix 2 up to and including the end of the Period during which the day named in the Statement of Practical Completion occurs.
   • thereafter Application for Interim Payment shall be made as and when further amounts are due to the Contractor, and . . .
   • after the expiration of the Defects Liability Period named in Appendix 1, or . . .
   • on the issue of the Notice of Completion of Making Good Defects (whichever is the later) . . .

- provided that the Employer shall not be required to make any Interim Payment within one calendar month of having made a previous Interim Payment.

30.3.2 Each Application for Interim Payment shall be accompanied by such details as may be stated in the Employer's Requirements.

30.3.3 Subject to Clause 30.3.4 the Employer shall pay the amount stated as due in the Application for Interim Payment within 14 days of the issue of each Application for Interim Payment.

30.3.4 If on receipt of any Application for Interim Payment the Employer considers that the amount stated as due in the Application is not in accordance with this Contract he shall forthwith issue to the Contractor a notice with reasons to that effect and shall pay at the same time as the issue of that notice such amount as he considers to be properly due as an Interim Payment.

30.3.5 The payment by the Employer of the amount referred to in Clause 30.3.4 shall be without prejudice to the rights of the Contractor in respect of any amount which he considers has been improperly withheld by the Employer or in respect of any payment by the Employer which he considers was not in accordance with this Contract."

Clause 30.4 then deals with the levels and mechanics of retention, which mirrors the JCT 80 provisions, but with the drafting amended to cover the Alternative A and B methods specific to JCT 81. In practice the above system, i.e. "You send me an invoice once a month and so long as it is reasonable I will pay within 14 days. If not, I will let you know what I do not like and pay you what I do agree" can and does work—on the simpler, lesser value, design and build projects.

For the more complex or bigger value design and build projects such a simplistic method based on trust between the parties is fraught with dangers. Whilst the managing directors in the employer and contractor organisations may well be able to telephone one another on first-name terms, at site level working relationships can and do go wrong from time to time.

Typically the contractor's surveyor has to prove himself to his superiors by making his target profit, and assuming there are no mistakes, either way, in the tender price which the site surveyor inherits from his estimating department via the director's office, he can only make his target by maintaining the buying margins, i.e. on subcontract placements and on main materials identified in the tender.

However, contracting being an imperfect science, full of inefficiencies and risks, the site surveyor will usually set about trying to build up a further contingency margin by re-bidding various packages or trying to beat preferred subcontractors down.

On the other hand, when it comes to the first interim valuation and subsequent interim valuations, the contractor's site surveyor will naturally seek to maximise the return to the contractor, his immediate employer. A balancing act must therefore be performed by the contractor's site surveyor and at the end of the day he cannot claim money twice over—some do, but that is another story!

The above comments are of course equally applicable to most forms of contract, but the differences between JCT 80 and JCT 81 come into play with the absence of a third-party certifier.

Under JCT 81 the design and build contractor's application is delivered or posted direct to the employer, who will usually be a corporate body, with or without in-house individuals with knowledge of the building industry. In practice the contractor's application will therefore be forwarded (if not sent direct) to the in-house individual, or out-of-house consultant, named in the contract as the employer's agent.

However, the employer's JCT 81 contractual duty is to deal with the contractor's interim payment application within 14 days, and to actually pay all, or to pay part with stated reasons for not paying the whole. It is therefore not unknown for some design and build contractors to rely on the above provisions, word for word, and deliberately ambush unsuspecting employers.

It can happen and has happened that a design and build contractor has made application direct to the employer's agent and that for whatever reason the employer's agent has failed to inform the employer of the existence of an application. As such the employer could have no knowledge of such application and no opportunity to make the necessary payment.

There is then nothing in the JCT 81 contract to prevent the contractor making an Order 14 application to the High Court, seeking summary judgment for payment of all monies claimed, even including a wholly unsubstantiated loss and expense claim on the basis that the employer has failed to comply with his obligations at Clause 30.3.4, i.e. no notice of objection to the contractor's payment application and/or no reasons stated for not paying any amounts claimed within 14 days.

Such tactics are thankfully not typical of design and build, but be warned —they can happen, and if they do they can be very time-consuming and expensive in legal and other fees to defend. In such circumstances, if you are the aggrieved design and build contractor, I would suggest there are three remedies, short of litigation, to obtain payment:

(a) Common sense—pick up the telephone and enquire whether there is any reason you have not been paid.

(b) Contractual remedy No. 1—send a seven-day letter by recorded delivery and, if still not paid after the said seven days, stop work, sending another letter recorded delivery under the determination provisions of Clause 28.1 confirming the fact of continued non-payment *and* actually stop all works.

(c) Contractual remedy No. 2—assuming you believe there is a dispute—seek adjudication or arbitration, subject to the provision of your particular contract. Whether simple failure by the employer to pay is a "dispute" or a breach, or both, I leave to my legal friends, but one can only adjudicate or arbitrate a "dispute".

Payment dispute avoidance

The foregoing example of contract abuse demonstrates how essential it is that any design and build contract is properly set up before tender. Accordingly, in most projects the interim payment procedure is elaborated by specific provision in the contract particulars section of the employer's requirements. If such elaboration is intended, the signed contract should be amended at Clause 30 to incorporate a reference to the employer's requirements, initialled by both parties.

In the case of Alternative B: Periodic Payment projects, such elaboration can take the form of providing that:

(a) The contractor submits his draft application direct to the employer's agent, on a regular date each month;

(b) Within three working days the employer's agent and the contractor's surveyor meet on site and agree items and amounts payable;

(c) The contractor then formally submits his application to the employer, copy to the employer's agent—*from which time the 14–day contractual period will run,* unless some other payment period has been entered in the contract.

Thus the employer cannot be ambushed or otherwise taken by surprise, and it remains for the employer's agent to confirm to the employer that the net sum claimed by the contractor on the interim payment application is indeed in order and should be paid within the 14 days (or such other period).

For the avoidance of doubt regarding payment and other periods I like to see a clause at the front of the specification or bills of quantities which states that unless otherwise stated all references to days are calendar days, thus 14 days equals two weeks, but three working days means what it says.

If nothing else, such elaboration of the standard JCT 81 interim valuation procedure makes good sense for the business efficiency of any project, as few employers have in-house capability or capacity for dealing direct with such matters. In any event even fewer have substantial funds on current account, so they do need notice of necessary withdrawal dates and likely amounts required from other accounts or funding sources.

Inevitably there are certain safeguard clauses which are useful on behalf of the employer and should be put into any design and build contract to give power to the employer or employer's agent to enquire into details of the contractor's interim payment application, e.g. proof of payment to subcontractors.

Some of these safeguard clauses are in the nature of first-phase defences, e.g. no bond, no first valuation and are dealt with in greater detail in Chapter 3 in "Contractor's Obligations and Risks". Some are by way of "catch-alls", e.g. power for the employer to appoint an auditor and are discussed at 6.4 in "Final Account".

On most design and build projects such employer-protection clauses are not needed in practice, but in this competitive, commercial world one has to legislate for the contractor who will not play by the usually accepted rules —whether bid low (and sometimes this is deliberate company policy) or simply because of a policy of maximising profit at all costs.

Finally, this chapter on interim payments would not be complete without stating the obvious: if you fail to lay down the ground rules before the first valuation under Clause 30.3.2 you will only have yourself to blame when the contractor makes interim applications in unparticularised form.

Whether employer or employer's agent, you will be disadvantaged in any subsequent negotiation with the contractor and this area can be the origin of a festering dispute.

Clause 30.3.2 requires that : "Each Application shall be accompanied by such details as may be stated in the Employer's Requirements." It is therefore recommended good practice to prescribe the following interim valuation "ground rules':

1. Each interim application for payment (if periodic payments are applicable) shall be in the form of the contract sum analysis and shall show against each item within each element of the works:

   (a) The total quantity, unit rate, and cost allowed in the accepted tender.

   (b) The quantity now built, or percentage of the whole, extended by the contract rate, to show the current value of each item within each element of the works.

2. In respect of changes (Cl. 12.2) a breakdown of the value claimed on account shall be given, either by quantity executed and unit price, or by apportionment of lump sum value, whichever is appropriate, subject to 6 below.

3. Value claimed in respect of provisional sum expenditure (Cl. 12.3) shall be supported by any invoices as may be requested by the employer or his agent prior to approval of the amount claimed and before admission to the final account.

4. Value claimed in respect of materials and goods shall be similarly capable of being supported by production of invoices as may be requested by the employer or his agent prior to approval of the amount claimed.

5. Design fees will only be paid subject to prior written agreement with the employer which should be in the form of a schedule related to given project dates, e.g. month 1, month 6, etc.

6. Only changes (Cl. 12) agreed in writing by the employer as to *cost* and *time*, or instructed to be put in hand without prior agreement, will be admissible to interim valuations. Claimed changes which do not so

comply will only be payable on final account at the employer's discretion, without entitlement to interest thereon for the contractor.

The JCT 81 Supplementary Provisions

This last "Ground rule" is deliberately draconian and goes to the heart of the design and build philosophy as discussed in Chapter 3.3 in "Certainty of Cost and Time". This is intended not to be a unilateral benefit in favour of the employer, but a mutual benefit in favour of *both* employer *and* contractor. If put into practice, it snuffs out potential disputes at birth and makes the final account process a rolling account process rather than an after-the-event battle-ground: see 6.4 in this chapter.

This provision is in fact to be found in the Supplementary Provisions at Clause S6, so if your contract incorporates the Supplementary Provisions by the correct deletion of the alternatives offered in JCT 81 Appendix 1, Article 1 the position should be secure.

I say "should" with good reason. It is no good if, as employer's agent, you then turn a blind eye to actually enforcing the Supplementary Provisions. One employer's agent managed to receive and approve successive interim payment applications from the design and build contractor, and despite the Supplementary Provisions being incorporated in the contract, allowed substantial amounts for a long list of claimed changes which had not been pre-costed and agreed between the parties.

When the employer had the project audited, and on advice resisted further payments, the contractor ran a litigation on the basis of deemed acceptance of such changes, and the employer's agent's professional indemnity insurer was then not best amused to be joined in as third party by the employer.

Payment provisions need to be workable

Such are some of the perils and pitfalls of the interim payment process—all too many individuals in the industry still do not appreciate the very significant distinction between JCT 80 and JCT 81, and this includes not just well-intentioned architects, quantity surveyors or engineers who accept appointments as employer's agents. It also includes design and build contractors, whose staff particularly at site level may have been recruited or relocated for a particular JCT 81 project, but who were most recently employed on a JCT 80 contract.

A JCT 81 project requires a different mental approach, and if both party's site-level representatives appreciate that requirement, then at least the project should get off on the right footing.

Unfortunately all too many contracts drafted to date, particularly those with heavy amendments to standard forms, suggest lack of understanding of what actually happens, or should happen, at site level in respect of persuading one party not to seek undue payment, and persuading the other party to actually pay.

Hopefully the foregoing advice, from one who has been called in to troubleshoot problem design and build projects where the interim valuation process has failed, will throw some light on the subject. Hopefully too it may actually help prevent future design and build abuses and payment disputes by encouraging proper drafting of interim valuation "Ground Rules".

6.3 VARIATIONS

Any project, whether procured on a standard JCT contract or other form, will be liable to variations ranging from changes of mind on the part of the employer or to unforeseen problems raised by the contractor and/or his subcontractors.

As has been explained in Chapter 4, under JCT 81 variations are defined at Clause 12 as "changes in the employer's requirements and provisional sums", or more simply "Changes".

Hopefully the reader will now appreciate that under JCT 81 it is for the design and build contractor to establish that the "Change" is indeed attributable to the employer changing the content of the employer's requirements as opposed to the employer merely giving the nod, so to speak, to a development or refinement of design detail suggested, apparently gratuitously, by the contractor.

As a "good practice" tip I would therefore suggest two guiding principles in identifying true Clause 12 changes:

1. Is the change as contended by the contractor functional or cosmetic?
2. Was the employer aware at the time he agreed to the intended change that the original bargain, as per the contract sum or the agreed *time*, was being varied?

It also bears repeating that the concept of JCT 81 does not envisage the usual glut of JCT 80 variations; the employer should limit himself to essential matters only, apart from deciding on reserved matters such as approval of materials against a stated prime cost sum or the expenditure of provisional sums.

The reverse of the coin is that the design and build contractor must take a broader brush approach, accepting the rough with the smooth as *his* design evolves and conflicts have to be resolved without recourse to the employer under Clause 12.

If one opts for the "Hard" version of JCT 81 offered by the Supplementary Provisions and Clauses S1 to S7, then changes should not be a problem area—or at worst, something that a quick and cheap reference to adjudication cannot resolve.

6.4 FINAL ACCOUNT

As any final account is essentially a function of the contract sum, plus or minus variations, or in JCT 81 language "changes", it follows that, if the variation or change-order procedure is defined in the contract documents, *and* followed during the course of the contract, the final account will fall into place relatively easily.

Again, the "hard" version of JCT 81 offered by the Supplementary Provisions and Clauses S1 to S7 with the concept of agreed pre-costing and agreed *time* implications of changes, if applicable, is as near to a perfect dispute avoidance formula as one could hope for.

When the JCT 81 Supplementary Provisions, or a derivative thereof, are incorporated in a design and build contract, the final account is a rolling process, i.e. the goalpost of *cost* is moved by pre-agreement and both parties know where the cost goalpost is at any one time. Unfortunately, many a JCT 81 design and build contract has been let without the option of the Supplementary Provisions being incorporated—mainly, I suspect, out of ignorance by employer's representatives, and including, I am afraid to say, some solicitors.

This is probably due to the fact that the "Lord's Prayer" so to speak is to be found at the very back of the Hymn Book, i.e. the Supplementary Provisions do not appear until page 64 of the JCT Standard Form of Building Contract With Contractor's Design 1981 Edition.

If the Supplementary Provisions option has not been chosen the design and build contract is what I have dubbed the "soft" version, and the final account procedure can be akin to the opening of the hunting season, i.e. colourful but unnecessary and unpleasant. If you do have to get involved in such a pursuit it might be helpful to be familiar with the particular rules. JCT 81 Clause 30.5 can be summarised as follows:

30.5.1 The contractor prepares the final account and submits to the employer within three months of practical completion including " . . . such supporting documents as the Employer may reasonably require".

30.5.2 Provides for adjustment of the contract sum and refers to Clause 30.5.3 for the mechanics of so doing.

30.5.3 Deals in detail with allowable additions and deductions to the contract sum, there being nothing which is peculiar to design and build, i.e. it is all thoroughly recognisable from JCT 80.

30.5.4 Formally provides for a final statement setting out the contractor's contended final account total, versus monies already paid by the employer.

30.5.5 Provides for when the contractor's contended final account becomes conclusive, i.e. the employer must object, if he so wishes, within *one month of the later of three events*:

- The end of the defects liability period, or
- The day named in the notice of completion of making good defects, or
- The date of submission of the final account and the final statement to the employer by the contractor.

So far, so good—the employer and his agent have one month to ask for further details of the contractor's contended final account and to take issue with them before the contractor can enforce payment.

Of course the "don't-want-to-pay" brigade of employers can then raise a smokescreen defence by asking for all sorts of details, and continuing to do so, provided they actually record their non-acceptance of the contractor's final account.

It might have been happier if the JCT drafting committee had added at the end of Clause 30.5.1 " . . . prior to Practical Completion", i.e. making the employer inform the contractor as to the level of detail required. It might also have been helpful if the JCT drafting committee had for the avoidance of doubt repeated the interim valuation procedure in that the employer should pay over further agreed monies, and only withhold disputed monies.

The final account provisions of JCT 81 Clause 30 then get more interesting: under Clause 30.5.6, if the contractor fails to submit his final account and final statement to the employer within the prescribed three months of practical completion he loses the initiative. The employer may then write to the contractor putting him on notice that the final account and final settlement must be produced within a further two months of the employer's letter, failing which the employer himself will prepare the final account.

This is a remarkably powerful provision and one that I have used myself. It so happened that the contractor's three-month default period ended on 7 November, so my letter as employer's agent dated 10 November required the contractor to produce the outstanding final account by 10 January. Of course no contractor's quantity surveyor willingly foregoes his $2\frac{1}{2}$ week Christmas and New Year holiday, so it was all hands to the pump and a 22 December final account delivery.

The final account provisions of JCT 81 Clause 30 then get even tougher for the dilatory design and build contractor:

30.5.7 The employer's version of the final account can only be as good as " . . . the information in his possession . . . ".

30.5.8 The timescale for the employer to send his version of the final account to the contractor, assuming we are now at practical completion, plus two, plus three months, is the *later* of:

- The end of the defects lability period—usually month 12, or
- The day named in the Notice of Completion of Making Good Defects.

As the issue of the Notice of Completion of Making Good Defects is not a third-party umpire function as under JCT 80 it is entirely within the discretion of the employer under JCT 81. The design and build contractor who does not submit a reasonably detailed final account within 5 months of practical completion only has himself to blame if the employer, rightly or wrongly, then keeps him out of his final account monies for months, if not years. This can be done by the simple expedient of fault-finding in the works as built or even in the design, and by the employer not issuing the certificate of making good defects.

So we come to the final scene of this final account drama—what happens if the employer has the power to issue the final account, due to contractor default under Clause 30.5.6, and after taking his time actually issues his own version of the final account?

It is of course odds-on that the contractor will find both the detail and the level of the employer's final account unacceptable, but what can he do under the contract and what is the time limitation? Clause 30.5.8 provides that the employer's version of the final account is conclusive *unless* " . . . the Contractor disputes anything in that Employer's Final Account or Final Statement before the date on which, but for the disputed matters, the balance would be conclusive".

The answer to the second part of the question is then provided: under Clause 30.6 the contractor has 28 days after issue by the employer of the final account in which to object, presumably in writing, but again the JCT drafting committee has not been definitive, i.e. will a verbal objection suffice? On such an important point, I would like to think not, but it might be very expensive if one had to go to court to find out!

Finally, JCT 81 Clauses 30.8 and 30.9 provide for:

- Certainty as to the effect of the final account and final statement, if unchallenged, in respect of materials, goods or standards expressly reserved for approval by the employer;
- Procedures if arbitrations have been commenced but not proceeded with for 12 months;
- The design element being a continuing risk for the contractor irrespective of the issue of a final certificate.

6.5 LOSS AND EXPENSE

Loss and expense claims by the contractor under JCT 81 citing clause 12 changes should be the exception rather than the rule that they are tending to become under JCT 80. However, this does depend upon the JCT 81 contract provisions being properly set up and then being properly administered by the employer's agent.

The primary reason for this is that, as previously explained, it should be the contractor, not the employer, who is in the driving seat of JCT 81 projects in terms of design development and timely release of drawings to the employer for approval and comment *before* these are issued by the contractor "For Construction". Accordingly there should be few grounds for the contractor alleging the employer has delayed the project by late release of previously requested drawn information or issuing an undue number of variations, the two old chestnuts from JCT 80.

JCT 81 Clause 26 provides the usual "List of Matters" upon which a design and build contractor is entitled to found any such loss and expense claim, namely:

26.2.1   Opening up for inspection of any work or testing, subject to the examination or test proving satisfactory;

26.2.2   The late receipt of permissions or approvals related to development control matters;

26.2.3.1 Concurrent direct works by the employer, or the failure of the employer to carry out such concurrent work as promised and relied on by the contractor;

26.2.3.2 The failure of the employer to supply promised materials and goods;

26.2.4   Postponement of work by the employer;

26.2.5   Failure by the employer to give full possession of the site by the due date;

26.2.6   The issue by the employer under Clause 12 of instructions effecting a change or provisional sum expenditure;

26.2.7   The failure of the employer " . . . to give in due time necessary instructions, decisions information or consents . . . which the Employer is obliged to provide or give under the Conditions including a decision under 2.4.2".

Clause 2.4.2 refers to discrepancies within documents and there is also the usual proviso that the contractor should apply in writing, not too soon, not too late, for such instructions etc.

The important point here is that only instructions " . . . which the Employer is obliged to provide or give under the Conditions" qualify as a listed matter, i.e. any delay in design development approval by the employer is *not* a ground for the design and build contractor claiming loss and expense, as the underlying principle is that he, not the employer, is responsible for complying with the employer's requirements and the contractor's proposals. So long as the unapproved design development proposal is within the parameters so defined he carries on and builds. However, in practice any important design matter or key material selection should be put to the employer a second time on the basis: "If you do not object within . . . days we will proceed."

The above is therefore another distinct difference in emphasis as between JCT 80 and JCT 81, i.e. the onus is on the contractor, not the employer, to finalise drawn information in sufficient time to service the construction programme.

Clause 26.2.6, Changes in the Employer's Requirements, still remains the favourite hook for design and build contractors to hang loss and expense claims on—but if the employer opts for the Supplementary Provisions and Clause S6 at the outset of the project the problem should not arise, unless the contractor can make a cumulative disruption case under Clause S7.

If, however, a design and build contractor does submit a loss and expense claim under Clause 26 the same principles of evaluation still apply:

- Notification in principle to the employer as soon as the contractor is aware of a potential delay or loss and expense situation, together with best estimate of *cost* and *time*;
- Detailed notification when *cost* and *time* can be firmed up;
- If no lump sum or extension of time is agreed, causation, i.e. effect on critical path and actual loss, as opposed to notional loss, will need to be reasonably proven, preferably on an "open book" basis.

6.6 PROVISIONAL SUMS

Provisional sums have been previously discussed in Chapter 3 in the context of how uncertainties in scope of work at the time of tender can best be met by either the employer or the design and build contractor.

Apart from the employer's contingencies, which is a pure and sensible financial provision to guard against unforeseen project costs, every effort should be made to avoid provisional sums. However, where the required works cannot be adequately described in performance terms at time of tender the objective should be to at least describe the scope of work to be covered by a provisional sum.

By way of example, a provisional sum of £20,000 for reception desks would be more meaningful in contractor's programming terms if at least the employer says "5 No. reception desks" and gives some idea of where they are required and what size they might be, assuming the design is that far developed. Whether they are to be mahogany or veneered plywood can be resolved later by discussion and selective tender.

As regards final accounting for provisional sums the same principles as under JCT 80 apply to JCT 81—all costs should be "open book" with a sensible mark-up for the design and build contractor's overheads and profit.

Design costs involved in procurement of provisional sum work is separately discussed later in this chapter.

On the principle that authorised provisional sum expenditure is paid by the employer, at net cost, before the agreed uplift for overheads are profit, the design and build contractor who persuades the employer to firm up his brief, sometimes by way of good-looking samples, and then tenders the work, keeping trade discounts for himself, is improperly trading in provisional sums and as such is thieving from the employer.

On one project the design and build contractor, having entered fire fighting equipment as a provisional sum in the contract sum analysis, eventually produced an invoice for fire extinguishers, showing the quantity and unit rate column only, but with the adjoining column and the total blanked out. It was a simple task to telephone the company supplying the fire extinguishers using the telephone number at the top of the invoice. The hidden trade discount turned out to be 50 per cent!

6.7 DESIGN COSTS

As was made clear earlier when discussing the contract sum analysis document as required under JCT 81 it is important to distinguish between pre-contract and post-contract design costs which the contractor will incur. Not only is it important to be satisfied that such post-contract design costs are not too high but it is probably even more important that they are not too low. In other words, has the design and build contractor allowed sufficient resource for the properly considered design development work which will be essential if the project is to meet the employer's expectations?

At the best of times this will be a difficult judgment, but then what about changes under Clause 12 and the knock-on redesign these will inevitably incur? Whereas the finished work might show no significant cost difference, it is entirely possible that the proper consideration, particularly if health and safety matters are involved, of design implications may involve several hours work. In such a situation the consultant architect, or in-house draughtsman, will have raised a time record for his work and this will find its way eventually into the contractor's costing of the alleged change.

Hopefully sweet reason will prevail, but again it is better to be prepared and to have agreed the equivalent of a dayworks system, with presentation of such timecards at the end of each week, and to have agreed hourly rates for draughtsmen etc. However, such timecards must fairly differentiate between normal design development time and actual time directly attributable to the Clause 12 change requested or explicitly authorised by the employer.

Again, the operation of the Supplementary Provisions with the pre-costing requirement of S6, which includes design costs, will provide certainty for all concerned and is accordingly to be recommended as best practice.

6.8   TYPICAL PROBLEMS

Problem A

This concerned a JCT 81 design and build tender:

- The contractor submitted contractor's proposals consisting of:
  - (a) 35 pages of text,
  - (b) Employer's retitled plans, elevations and sections.
- Post letter of intent issue, but before any post-contract design work was commenced, the employer wished to revisit the principle of space utilisation, calling for substantive changes to concept and functions, but within the same four-storey building footprint.
- A formal Instruction was issued and the contractor remeasured all internal walls and partitions within the external walls, some of which had changed in specification, using the agreed revised drawings, and then valued them by reference to the contract sum analysis back-up information, i.e. mini-bill, to get to the new "Additions" value.
- The contractor then remeasured all the original internal walls and partitions within the external walls from the original drawings, costed them out to get to the "Omissions" value, contending a net extra cost on the basis of the "Additions" value exceeding the "Omissions" value.
- The employer's agent agreed the contractor's "Additions" remeasure and value, after the usual minor glitches, but disagreed the contractor's "Omissions" value on the basis that such was shown in the contract sum analysis as lump sums against the relevant sub-elements. When these five lump sum items were totalled they showed a value significantly higher than the contractor's measured omissions. Further, as they exceeded the agreed "Additions" value the employer's agent contended a net saving.

Subject to the different approaches as to how to value the "Omissions", both were correct in terms of measurement and mathematical extensions, and the dispute was worth some £35,000: who was right? Does one remeasure "Omissions" from original drawings, having to make assumptions as to specification, thicknesses and heights of walls where not detailed, or does one deal in lump sums as declared in the contract sum analysis?

When examined in lower level detail it was found that if one took the contractor's measured "Omissions" quantities away from his declared quantities in the contract sum analysis back-up document, or mini-bill, some 2,000 sq. m. of various types of partitions remained! No doubt the contractor knew this for himself, so had opted for his alternative approach. So the deeper question arose—under a JCT 81 design and build tender, is the contractor entitled to retain any "green" quantities which he may have inserted against oversights elsewhere, if that particular element or sub-element is substantially disturbed by a Clause 12 change?

Of course the contractor contended that the employer could not "cherry-pick" but in this particular case the Clause 12 change arose owing to a wholesale re-planning of all four floors by the employer. Had it been a loc-alised alteration of, say, one wing on one floor, a measured "Additions" and "Omissions" basis would have been the appropriate solution.

Reference was then made to the wording of Clauses 12.5.1 and 12.5.2:

> 12.5.1 "The valuation of additional or substituted work shall be consistent with the values of work of a similar character set out in the Contract Sum Analysis making due allowance for any change in the conditions under which the work is carried out and/or any significant change in quantity of the work so set out. Where there is no work of a similar character set out in the Contract Sum Analysis a fair valuation shall be made."
>
> 12.5.2 "The valuation of the omission of work shall be in accordance with the values in the Contract Sum Analysis."

The first issue was therefore: was the change "additional work", "substituted work" or "the omission of work"?

The employer's agent argued that the primary rule was reference to the contract sum analysis which by its elemental and sub-elemental lump sum nature enabled the "Omissions" value to be determined, but which did not permit the assessment of a remeasured "Additions" value without reference to the contract sum analysis back-up document or mini-bill. By using this second point of reference, the employer's agent had ascertained a fair "Additions" valuation and satisfied the precise requirements of Clause 12.5.1 in respect of the "Omissions" value.

The contractor on the other hand saw the contract sum analysis back-up document not as a mini-bill but only as a schedule of rates. The contractor then developed a very ingenious alternative argument. Having accepted the employer's agent's contention that the extended value (C) of each item in the mini-bill was inviolate, being an integral part of the agreed contract sum, he argued that the proper way of resolving the problem was to reduce the quantities (A) down to what they should have been had he measured the original drawing correctly at time of tender and to make an upward compensating adjustment to the unit rate (B) such that the lump sum value (C) was preserved. If he had succeeded in this suggestion he would have gained a double benefit—he would have recovered additional quantities, as against the *corrected* original quantities, *and* he would have recovered at the conversely corrected *higher* unit rate!

The contractor had then compounded the argument to a £100K-plus dispute by remeasuring all wall finishings, both "Additions" and "Omissions" as part of the same exercise by "anding-on" finishings items to the walls and partitions items, but somehow forgot that not all walls had plaster and paint both sides. In the original scheme as tendered, many walls had expensive tiling and wallpapering one or both sides, but on the contractor's chosen method of final account adjustment the employer was only given credit for plaster and paint!

Full marks to the contractor for trying—and being trying, but it goes to show what problems lie in wait for the unwary employer's agent.

Problem B

This involved the same project, same scenario—but how to adjust the final account following the same wholesale preconstruction layout change, but in respect of sanitary fittings (and all the associated plumbing work)? Now, one might think, what could be simpler than counting toilets and sorting out how many should be paid for? How wrong can one be?

So, taking just toilets, i.e. w.c. pans, and ignoring all other sanitary fittings such as basins, sinks and urinals, etc. and the associated plumbing, the facts were as follows:

(a) Valued declared in the contract sum analysis — Not available, being part of lump sum value for the sub-element

(b) Toilets and value declared in the mini-bill (not a contract document, but called for in the employer's requirements): 64 No. @ £110 = £7,040

(c) Toilets identifiable from floor plans included in contractor's proposals: 47 No.

(d) Additional toilets and value claimed on final account: 11 No. @ £110 = £1,210

(e) Total number of toilets installed in finished project: 56 No.

On the basis of original contract value, plus adjustments thereto, the contractor was contending total payment as follows:

(b) Value as in tender	£7,040
(d) Additional value	£1,210
Adjusted value payable	£8,250

On the other hand the employer's agent contended two options:

Option 1

(b)	Value as in tender		£7,040
(b)	Original number of toilets		
	declared in mini-bill:	64	
(e)	Revised number as built:	56	
		──	
	∴ A saving of	8 toilets	
		x £110	£(880)
	Adjusted value payable		£6,160

Option 2

(b)	Value as in tender		£7,040
(c)	Original number of toilets		
	shown on plans in contract-		
	or's proposals:	47	
(e)	Revised number as built:	56	
		──	
	∴ An addition of	9 toilets	
		x £110	£990
	Adjusted value payable		£8,030

Thus there were three possible answers to the apparently simple question: £6,160, £8,030 or £8,250. Which is correct?

The critical underlying question was: How many toilets were contractually deemed to have been included in the agreed contract sum, i.e. what was the base-line for adjustment purposes? The options were:

(i) The undisclosed number toilets supposedly contained in the design and build contractor's subcontractor's original scheme appraisal and which allegedly was the base-line for the 11 No. additional toilets now claimed; or

   (ii) The number of toilets declared in the mini-bill (64 No.), albeit the mini-bill was not a contract document in its own right and therefore strictly outside the Clause 12 requirement of value adjustment by reference to the contract sum analysis; or

   (iii) The number showing on the four floor plans (47 No.) as included in the employer's requirements and as re-issued in the contractor's proposals, albeit with the contractor's logo at the bottom right.

There was no evidence as to date or drawn basis of the subcontractor's original scheme appraisal and claimed adjustment thereto, so the contractor's submission contending for the equivalent of 75 toilets (£8,250 ÷ 110) was not seriously considered; but what about the employer's agent's options (1) or (2)?

- On the basis that Clause 12.5.1 requires adjustment by value, not by remeasurement *per se* and that no value of toilets alone was available by reference to the contract sum analysis, could either party rely on the mini-bill under the "fair valuation" rule?
- If one did refer to the mini-bill, could the contractor contend that the employer's agent was only entitled to refer to the unit rate column, and not to the declared quantities?
- If, given the contractor's mini-bill, value (C) was spoken for, and the contractor conceded that the unit rate (B) was contractually relevant, how could the contractor sensibly contend that the quantities (A) were not also contractually relevant?

The employer's agent ran the argument that, given the different numbers of toilets as between the contractor's proposals and the contractor's contract sum analysis back-up document, i.e. the mini-bill, then the situation was:

1. Tantamount to a discrepancy within the contractor's documents and therefore the employer could choose between the 47 No. toilets showing on the agreed tender drawings and the 64 No. toilets declared in the contractor's mini-bill—in which case the base-line would be 64 No. toilets.

2. Alternatively, it was clear that the contractor had allowed for 64 No. toilets in the accepted contract sum, i.e. more than showing on the agreed tender drawings, but that 56 No. toilets had been installed. As such, no additional value was justified as in (1), above, but arguably it would be wrong to claim a credit for 8 No. toilets as the change had not been specifically instructed by the employer, but had arisen due to the contractor's design development.

The problem did not stop there, as apart from the question of cold water supplies and drainage to toilets, the contractor was also claiming substantial additional costs in respect of toilet cubicles and passive infra-red (PIR) controlled ventilation—all on the basis there were 11 No. additional w.c.s, which was simply incorrect whichever base-line applied!

So be warned—it is essential not only that the contractual status of the contract sum analysis *and* the necessary back-up detailing is agreed prior to start on site, but also that the contractual status of the contractor's declared quantities is not in doubt.

CHAPTER 7

TIME ISSUES

7.1 PROGRAMME

Given that under JCT 81 changes should be far less numerous than variations under JCT 80 it follows that the circumstances of the design and build contractor having grounds to complain of late information release, multiplicity of changes, etc., should be relatively limited.

If the Supplementary Provisions option of JCT 81 has been written into the contract by the employer then the usual Clause 25 provisions as to the contractor giving written notice to the employer of:

- The relevant event relied on under Clause 25.4, and
- The expected effects thereof under Clause 25.2.2.1, and
- The estimated delay involved and the extended completion date required under Clause 25.2.2.2.

are overtaken by Clause S6.3.4 which requires the contractor within 14 days of receiving what purports to be a Clause 12.4 instruction to " . . . submit to the Employer . . . the length of any extension of time required and the resultant change in the Completion Date" *before* carrying out the work instructed.

Clearly if the design and build contractor does proceed with the instruction in good faith and *then* discovers he has a problem with programme co-ordination then he is in a weak contractual position, and must rely on the fair-mindedness of the employer or his agent to accept that the problem could not reasonably have been foreseen, e.g. supply dates changed by a specialist manufacturer after the Clause 12 change instruction was given and accepted.

The overriding principle under JCT 81 is that once the contract has been signed, i.e. the outline design and the construction duration agreed, the design and build contractor must build out the project by the required date in the employer's requirements, on the information given, irrespective of any provision for the employer having the right to comment on or even reject the contractor's proposed design development drawings.

Sensibly the contractor should allow in his programme for:

- Initial design development prior to start on site;
- Identified construction activities with sufficient durations;
- Allowance for incorporating employer's provisional sums including contingency expenditure;
- Allowance for contractor's own risks;
- Allowance for the employer or his agent having limited opportunity to examine and comment upon the contractor's proposed design development details;
- Commissioning and handover.

JCT 81 requires only, in respect of time, that there is an agreed completion date, or in the case of phased works, partial and a final completion date.

It is therefore most useful, in anticipation of a possible extension of time claim, that the employer's requirements lay down the "ground rules". For example:

1. The successful tenderer will be required to table a detailed construction programme for agreement *prior to the first interim valuation.*
2. Such a programme will not be a contract document but will be regarded as a statement of intent identifying the detail and timing of any input required from the employer to enable the contractor to meet his obligations under the contract.
3. Such a programme shall show, or otherwise identify by attached text listing activity references, the contractor's preferred critical path and the relationship thereto of all dependent activities.
4. In the event of any Clause 12 change issued by the employer, any effect in respect of the *time* required to complete the works shall be evaluated against the agreed construction programme.
5. Such a construction programme shall be re-issued every three months by the contractor incorporating any agreed changes and recording the progress of the works as built.
6. Any "float" time declared by the contractor's programme is the property of the project and is a *time* contingency to be used for the agreed benefit of either party as circumstances may require.
7. The contractor's programme is deemed to make due allowance for all elements in the contract sum including provisional sums, which include employer's declared contingencies.

The concept of "float" time is a brain-teaser for the lawyers. If the employer's requirements are silent should "float" be declared in a contractor's construction programme, and if so, who owns it? If it is so declared, can the employer issue variations under JCT 80 or changes under JCT 81 and expect not to have to grant an extension of time? All good questions, but why not avoid them in the first place by laying down the "ground rules" as suggested?

What is probably common ground, whether under JCT 80 or JCT 81, is that if the contractor has declared his intentions by way of a construction programme, then the employer and his design team, or the employer's agent, impliedly undertake to perform their obligations such that the contractor has every opportunity of achieving his preferred programme and the key events identified therein. However, the only express contractual obligation on the contractor in respect of time is to achieve the completion date, unless intermediate contractual "milestones" are written into the contract.

7.2 PRACTICAL COMPLETION

As previously mentioned practical completion is a "grey area"—of uncertain in legal definition and limited case law. It is of course a trigger point under all construction contracts in respect of the ending of contractor's obligations and risks, e.g. insurance, and the assumption of such risks by the employer. It is also a trigger for obligations both ways in respect of the making good of defects and for the resolution of the final account.

Where the employer does not actually need the building, e.g. a speculative development but no tenant yet signed up, there might be a reluctance on the part of the employer to agree that the building is fully compliant and therefore should be taken over—thus protecting his cash flow in terms of retention release, liquidated and ascertained damages and insurance obligations.

On the other hand the more usual situation is that the employer is desperate to take over the building—often having further fitting-out and commissioning works to do before he can open for business—and the contractor is equally keen to hand over, thereby:

1. Getting another interim payment;
2. Getting release of 50 per cent of the retention fund;
3. Limiting or removing liability to liquidated and ascertained damages.

In *HW Nevill (Sunblest)* v. *William Press* (1981) 20 BLR 78 Judge Newey decided that the test of practical completion was not that everything down to the last detail was truly complete but that the contractor had fulfilled its obligations excepting only *de minimis* defects which could be made good at the employer's convenience during the next few months.

This legal definition of practical completion appears to amount to readiness for occupation by the employer—leaving the employer to be the judge of how much disruption he is prepared to suffer whilst the contractor rectifies agreed *de minimis* defects.

Of course this leaves considerable room for judgement as to what is "*de minimis*" and the employer's discretion is likely to be influenced by the urgency or otherwise of the situation, e.g. if he has no tenant yet secured, his interpretation of "*de minimis*" could well be fairly strict. On the other hand if he is

committed to a tenancy agreement and needs the rent, *"de minimis"* may be differently interpreted.

The legal definition of practical completion was usefully considered further by his Honour Judge Hicks in the case of *George Fischer Holding* v. *Davies Langdon and Everest and Others* (1994) ORB 775.

The facts were slightly unusual in that clause 23.1, Date of Possession, had been amended by the addition of the phrase "but not so as to exclude the employer" i.e. the employer remained in limited occupation or retained the right of access.

The date for completion came and went and was not extended, and although the employer started to use part of the premises before the due date for completion, the partial possession provisions of the contract clause 17 were not implemented. Instead, Davis Langdon and Everest, acting as employer's agent, had issued a "Substantial Completion Certificate" one month before the due date for completion—presumably to accord with the employer's gearing-up of his activities in the works area—and then issued a practical completion certificate almost two months after the due date for completion, but back-dated to the due date for completion: 14 April 1990. Unfortunately, the defectively designed roof had been letting in water over production stock and office areas from before the employer's agent granted his "substantial completion certificate", let alone from before practical completion as certified.

The Judge heard extensive legal arguments (which he covered in 12 pages of his decision), which can be summarised as follows:

- The usual provision is that any defects which appear during a specified defects liability period starting at the date of practical completion are to be specified by the responsible officer and made good by the contractor. *The implication is that defects already apparent, unless very minor, are inconsistent with the achievement of practical completion.*

The second sentence above is my emphasis and follows previous decisions in *J. Jarvis and Sons* v. *Westminster Corporation* [1970] 1 WLR 637, and *H.W. Neville (Sunblest)* v. *William Press* (1981) BLR 78. I have also taken the liberty of altering the clause references as quoted from the particular contract in this recent case to accord with the regular numbers of the JCT 1981 contract.

His Honour Judge Hicks then dealt with the not unusual situation of the employer moving in before the contractor moves out, MCL being the contractor in this case.

- "I do not see how the plaintiff can have made any significant use of the office block for some weeks after that at best and there was *no evidence that MCL (which still had a permanent force on site) was excluded from possession of any part of the premises or hindered in what it was doing to complete unfinished work and make good defects.*"

- "That being so, and having regard to the terms of Clause 23.1 of the contract, I see no basis for the suggestion that what happened on 14 April gave MCL any sort of claim to the issue of a certificate of practical completion in *flat contradiction of the true situation*."
- "I reject the plea of contributory negligence based on the alleged 'taking possession' both for the above reasons and because *it was for DLE to advise on such matters and there was no evidence that they did*."

Again, I have emphasised what I take to be the key phrases which point the way to future practice.

The last point is, in my view, particularly significant, i.e. the "duty to warn", given the pressures being exerted by the employer for occupation and the inevitable representations from the contractor for further payment, release of retention, the commencement of the defects liability period, and the ending of liability for the time overrun, i.e. liquidated and ascertained damages.

However, one of the two major issues in this case was the prejudice caused by DLE in certifying practical completion, in that it triggered the surrender of the performance bond. Given the major problem of a defectively designed roof over most of the works areas, the performance bond was the employer's only real protection under the contract.

Under pressure from the contractor, the employer sought DLE's advice on this point, sending DLE the bond document. No doubt mindful of the clear provisions of clause 17.1 concerning partial possession by the employer, DLE then decided that, practical completion having been granted, the contractor was entitled to the return of the bond and they themselves returned the bond to MCL.

His Honour Judge Hicks has therefore decided apparently on the facts of this case that JCT 81 clause 17 is of no avail to the contractor unless the contractor can show that all work has in fact reached practical completion based on the "*de minimis*" defects rule—and I emphasise "on the facts in this case", i.e. the employer never totally vacating some areas.

In DLE's defence, it has to be said they had issued a qualification document entitled "reserved matters" apparently at the same time as their "substantial completion certificate", but the items listed failed to include the known roof problems.

DLE did no better when it came to issuing practical completion—the contemporaneous minutes of meeting did not record any problems with roof leaks, so the three issues were:

1. *Had Practical Completion been achieved by 14 April 1990 or any other relevant date?*
   No, said the Judge.
2. *Whether the document issued by DLE on or about 6 June 1990 was the practical completion certificate required by clause 16.1 of the contract (whether or not properly issued)?*

Yes, said the Judge, in four pages in his decision, compounded by DLE's reliance on a compromise agreement allegedly authorised by the Employer.

3. *Whether the document (whether or not such a certificate) was properly issued as between employer and employer's agent, either under the contract or pursuant to some subsequent transaction?*

On this third issue as between employer and employer's agent—the bond having been surrendered—it came down to the alleged compromise agreement.

Such agreements, i.e. commercial decisions made under duress because the contract mechanism has broken down and the contractor is threatening litigation (with its high cost risk, not to mention uncertainty of result), are far from unusual.

However, as His Honour Judge Hicks decided, such a compromise agreement *must* be properly recorded if one party to that agreement subsequently seeks to rely on it in mitigation:

- "There is simply no evidence of a concluded agreement of that kind at any time before the issue of the document [practical completion]. It is inconceivable that a matter of such importance would not have been reduced to writing . . . ."
- "The plaintiff should have been advised, but was not, as to the merits and demerits of MCL's demands and the strengths and weaknesses of its own bargaining position . . . ."
- "If any agreement emerged DLE should have made a careful record of its terms and obtained confirmation by both parties of its accuracy."

To summarise, therefore, the legal position on practical completion appears now to be:

- Unless specifically recorded and reserved between "employer and contractor for making good during the defects liability period, *any defect known at the time practical completion is requested by the contractor must preclude the issue of such a certificate by the employer's agent.*
- *In the event that the employer requires to take possession after the due date for completion, but before practical completion can be certified, the contractor must allow the employer parallel occupation,* but the employer must not hinder the contractor in the completion of the required works—a situation usually referred to as occupation under licence—which of course should be confirmed in writing.
- The contractor's obligations under the contract in terms of insurances and any performance bond remain *until all the works are fully compliant and Practical Completion is certified, but subject to . . .*
- *Should the employer take possession and the contractor effectively vacate part of the works by agreement with the employer, a certificate of practical*

completion should be issued for that part—with the attendant release of retention on a pro-rata basis.

However, all of the above gives one clear message to those drafting contracts—for the avoidance of doubt why not write your own "ground rules" on a project-specific basis in respect of practical completion?

So long as you incorporate the basic legal principles discussed above and elaborate them only as necessary based on geographical description or operative need of your project, you should not go far wrong and could well avoid getting trapped into the sort of "compromise" situations that ensnared Davis Langdon and Everest.

Had the design and build contractor not gone into liquidation then this case would not have arisen—but then if the roof had been properly designed in the first place, or condemned by the employer's agent as patently defective in concept, then the contractor might not have needed to go into liquidation. A cynical view maybe, but the bottom line in design and build is that the employer is entitled to the delivery of a fully compliant building, fit for the stated purpose.

In the context of JCT 81 design and build, as opposed to JCT 80, there is no difference as regards the above *"de minimis"* defects test as to whether the works should be certified for practical completion or not, but I would submit there is potentially a critical difference in respect of mechanical and electrical services design liabilities. For example, under the JCT 80, when the employer grants practical completion he is saying in effect that he is satisfied that the contractor's mechanical and electrical design contractor has reasonably carried out his (the employer's) required mechanical and electrical design.

If the employer has set up the design and build contract properly he will have laid down a strict commissioning, testing and handover procedure to prove both the installation and the design—with structured stages and defined documentation—all of which must be cleared in good time *before* the final practical completion inspection. If something does not work on the day of practical completion which then involves the contractor in taking down ceilings in otherwise finished rooms. This can be disastrous.

Under JCT 81 the granting of practical completion by the employer is saying that he is satisfied that the *contractor's mechanical and electrical design*, as well as the installation, is fit for purpose. Any contractor's programme should therefore sensibly show services testing and commissioning as a distinct activity in its own right. This is particularly relevant on JCT 81 where the employer may have had concerns but no tangible reason to reject the contractor's proposed systems. In this situation the employer needs reassurance by way of seeing the services systems proven by performance on handover, and further proven over a full heating season.

Otherwise the only advice as to practical completion is that it is always easier to get perceived defects corrected as a precondition, before signing the works off and taking possession, rather than on a promissory basis after occupation, when the contractor usually has better things to do, like earning money on new work!

7.3   EXTENSION OF TIME

Clause 25.4 of JCT 81 sets out "Relevant Events" which are the basis for any entitlement which the contractor must cite and be able to establish if he wishes to make representation to the employer or his agent for the agreed completion date being extended.

The prescribed risks are in real terms no different under JCT 81 than under JCT 80, e.g. Clause 25.4.2, Exceptionally adverse weather conditions, and on the face of it, Clause 25.4.5, Compliance with the employer's instructions. However, as previously explained, on a properly administered JCT 81 one should see relatively few changes and accordingly contractor's claims for extension of time should be few and far between.

If the employer has taken the further insurance of inserting the Supplementary Provisions, then such claims have no standing—Clause S6 requiring pre-agreement before the work is undertaken.

Likewise the JCT 81 contractor notice provisions of Clause 25.2 and the employer's obligation to refix the completion date under Clause 25.3 are eminently recognisable from JCT 80, but of course are irrelevant if the Supplementary Provisions apply. The only further point of note is that, unlike under JCT 80, there is no obligation on the employer under JCT 81 to refix a new completion date in the event of no notice being received from the contractor.

Having now reviewed how the optional JCT Supplementary Provisions transform an inherently "soft" contract I can only surmise that it might be eminently sensible for all concerned if the JCT Drafting Committee at their next revision consider making the Supplementary Provisions compulsory unless deleted, as I do not believe any properly advised employer would wish to do without them.

This statement might be misconstrued as placing an undue risk on the design and build contractor, but there is really no downside for the contractor. On any project, certainty of *cost* and *time* should be a mutual benefit for both employer and contractor, reducing the opportunity for disputes and the attendant costs of litigation or arbitration. So why not have "hard" rules, with the safety valve of adjudication?

7.4   LIQUIDATED AND ASCERTAINED DAMAGES

The only point of interest particular to JCT 81 and liquidated and ascertained damages is the necessary rewording of the JCT 80 provision to allow for the differences in the final account procedure, i.e. contractor's final statement in the absence of an architect's final certificate under JCT 80 and the limitation of the employer's right to deduct damages for non-completion.

Otherwise the procedure is the same in principle—if the contractor fails to complete by the due date for completion the employer may serve written notice on the contractor to the effect that he will be withholding liquidated and ascertained damages.

The significance of JCT 81 is, again, that in the absence of a third-party umpire, it is of course the employer who decides whether or not to extend the completion date and the employer who then decides if practical completion can be granted.

The employer must therefore be careful not to abuse his double-edged power and may be well advised to call in a second opinion audit to take an independent view of the facts. Quite possibly this precaution will prevent the employer taking a false position, which the contractor has little option but to challenge by way of serving notice of arbitration or litigation.

7.5 TYPICAL PROBLEMS

Problem A

JCT 81 "soft" contract, i.e. Supplementary Provisions not incorporated.
Tender enquiry: 65 weeks construction period.
Lowest tender: 52 weeks offered.

At post-tender interview the employer's team pressed the design and build contractor as to whether 52 weeks from start on site to completion was too ambitious. The contractor was adamant and agreed to put the completed project on a care and maintenance, plus insurance basis, if completed before the employer was ready to take the building over at Week 65.

The project was then not properly controlled by the employer's agent, and the contractor at Week 63 submitted an extension of time application alleging cumulative effect of Clause 12 changes. Practical completion was granted at Week 75 and 10 weeks' liquidated damages were taken by the employer.

Given the above scenario the following questions arose:

1. Under JCT 81 the contractor is responsible for design and therefore the timely issue of drawn information, so can he reasonably claim to be overwhelmed by changes as late as Week 63 ex 65?

2. Given that the only contractual date was Week 65, could the employer reasonably ask the contractor to demonstrate that *but for* the alleged changes he would have completed by Week 52 (to which his accepted tender related)?

3. Could the employer contend that the reserved work value as covered by provisional sums and contingencies was deemed to be included in the contractor's declared 52-week construction programme?

4. In short, who owned the 13 weeks of "float" time?

My initial approach as auditor was that:

- The contractor had in principle to complete the agreed value within the agreed time, i.e. 65 weeks. Therefore the employer was entitled to omit the value of all provisional sums including contingencies and then have the benefit of changes up to the value of such omits without having to grant an extension of time.
- The contractor then had to show that any such change instruction took him by surprise *or* was patently late, after due warning that late information would have an influence on the critical path.
- The contractor then had to show how any *net* additional expenditure arose, e.g. not just more expensive equipment selection, but genuine additional scope of work, coupled with timing implications as above.

Problem B

JCT 81 "hard" contract, i.e. Supplementary Provisions incorporated.
Provisional sum of £3,500 for a feature chandelier to be suspended from soffit of the second floor slab and hung over the feature staircase leading up from the ground floor entrance lobby.

The design and build contractor did not identify this item as a key procurement event and only sought the employer's instructions as to the expenditure of this provisional sum with 12 weeks to go to practical completion.

A modern chandelier was selected from a standard trade catalogue and an order placed with 10 weeks' supply quoted. Seven weeks later the supplier advised problems with globe bulbs—a fire in the Italian factory allegedly. The employer's agent then negotiated alternative "globe" bulbs with the UK chandelier manufacturer, but the bulbs would not be at the factory for trial fitting for four weeks. Then the whole unit must be brought to site, hung, reassembled, etc. In the meantime two further weeks elapsed and the contractor then claimed eight weeks' extension of time, including taking down the ceiling, special scaffolding, protection of the glazed staircase balustrade, wiring back to fuse board, fire stopping, etc.

In reality the contractor was still painting and installing second fix electrics in 25 per cent of the building area and desperately needed an eight-week extension of time to avoid liquidated damages being deducted by the employer.

Counsel's opinion was sought and, despite the evidence that:

1. The contractor always knew the total weight of the chandelier and so had every opportunity to install the secondary steelwork necessary to the underside of the second floor soffit, but had failed to do so, and
2. The electrical subcontractor had anticipated the problem and had provided a separate spur and plug-in socket,

counsel advised, notwithstanding the substantial state of incompletion else-where, that an extension of time should be granted and liquidated damages should be returned, plus interest—seemingly on the basis that the employer's agent, having made a selection from the standard catalogues offered by the contractor, had then taken over responsibility for timely procurement when getting involved in negotiating alternative "globe" bulbs.

Problem C

JCT 81 "hard" contract, i.e. Supplementary Provisions incorporated.
PC price for wallpaper to be selected by employer.

Contractor offered a sample board with three months to go to practical completion. The employer selected within the PC price band, but two weeks later the contractor advised that the stockist held insufficient quantity of rolls with same batch number, and short of a special production run (premium price and many weeks) the contractor requested the employer to reselect, which was done immediately.

Problems then arose with the green paper hanging: the contractor blamed the subcontractor's quality of work, etc. when the paper shrunk and vertical gaps appeared. The employer's agent suggested back painting in green at vertical joint positions as a precaution and the subcontractor called in the wallpaper manufacturer's representative, who advised that the contractor had used a non-approved scrim jointing product on the dry lining which had caused a chemical reaction with the recommended wallpaper paste!

The contractor then submitted to the employer:

1. A claim for extra over cost of the alternatively selected wallpaper (additional costs not advised at time);
2. A claim for extension of time in reselection, re-ordering and difficulty of hanging;
3. A further *time* and *cost* claim, if you please, for having to order green paint and applying same!

As employer's agent my decision was negative, negative, negative—but was I necessarily correct? Certainly the contractor did not think so!

CHAPTER 8

QUALITY ISSUES

8.1 THE CLIENT'S BRIEF

In Chapter 3 we have seen how the whole procurement process commences with the client's brief, and with the strategic decision as to whether to go for a reasonably designed scheme at tender stage, or whether to leave the design requirements "light and fluffy" but within stated parameters. Indeed if the client is unable to work up a detailed design in the time available, i.e. he needs to have a contract in place before a financial year end, or he simply needs a "fast-track" project, he has essentially to choose between design and build and management contracting as his procurement strategy.

If one opts for management contracting, one only buys limited time before having to procure the works packages based on design by the architect—with the attendant risk that the design is still incomplete or late, giving grounds for time and cost claims from the management contractor.

If on the other hand, time is on the client's side he should not readily opt for design and build and JCT 81, in preference to an architect-led lump sum contract such as JCT 80, without being very sure that procedures are in place for:

- *Quality* control of the finished product;
- The reasonable incorporation by agreement of any changes the client may wish to introduce and their implications, i.e. *cost* and *time*, if applicable.

Hopefully the message of Chapter 3 has been received: the client opting for the design and build route gets one chance only to lay down the "Ground Rules", i.e. the employer's requirements document.

Not only should *quality* of materials and workmanship be specified by reference to national standards, but being design and build it is important to also provide for:

- Design criteria to be followed by the contractor in his design development;
- Supervision during the progress of the works by the employer's agent or other delegated professionals;

- Validation of the contractor's chosen design, especially structural engineering and mechanical and electrical installations;
- Independent verification of any commissioning tests carried out by the contractor;
- Contractual protection, i.e. design warranties locked into professional indemnity insurance in favour of the employer, should any link in the contractor-led design chain break or fail to deliver.

The importance of a fully considered employer's requirements document comprehensively setting out the client's brief and his terms for doing business cannot therefore be understated, and inherent in this is the *quality* of design and *quality* of the finished scheme.

8.2 DIVERGENCIES AND DISCREPANCIES

A JCT 81 design and build contract will consist of three base documents:

- The employer's requirements;
- The contractor's proposals;
- The contract sum analysis.

Each of these three base documents can in fact be a single document or a set of related documents—see JCT 81 Appendix 3. It follows that problems can arise if there is any conflict of information *within* each of these three, or as *between* any one and either of the other two.

JCT 81 distinguishes between "divergencies" and "discrepancies" by neatly limiting the former to problems of definition in the site boundaries as declared in the employer's requirements and the physical situation found on the ground at handover of the site by the employer to the contractor—Clauses 2.3 and 7. This therefore leaves the bulk of potential mismatches in the tender information and contract documentation as signed under the collective title of "discrepancies".

Unfortunately JCT 81 Clause 2.4.1 only deals with discrepancies *within* the employer's requirements, *and* where the contractor's proposals do not assist. Similarly, Clause 2.4.2 only deals with any discrepancies *within* the contractor's proposals.

Various "What if" scenarios immediately suggest themselves, e.g.:

1. What if there are discrepancies *between* one and another? Which document of the three base documents takes contractual precedence?
2. What if the contract sum analysis shows details as being included in the contract sum but which are not mentioned or conflict with details set out in the employer's requirements or the contractor's proposals?
3. What if the contract sum analysis suggests the contractor has not allowed for some element or detail required in the employer's requirements, but not expressly excluded in the contractor's proposals?

The importance of these questions is that, depending on the answers, Clause 12 changes will either kick in or not, and without Clause 12 kicking in there can be no contractual grounds for additional *cost* or *time*.

These questions must therefore be approached from a fundamentally different start point, as compared with JCT 80, i.e. under JCT 81 it is the contractor who has prepared two of the three base documents and who has taken responsibility for design detailing, which in turn dictates scope and quantity of the work required. Logic therefore says that the contractor's proposals and contract sum analysis are deemed to meet the totality of the employer's requirements, and therefore any discrepancy *within* either of the two contractor's documents, or as *between* the two documents, must be resolved on its merits, with the employer having the final decision.

The only exception to the above must be genuine "grey" areas within the employer's requirements which have reasonably caused misunderstanding and which reasonably were not picked up in the contractor's proposals. In this event, the benefit of doubt should be given to the contractor.

So what does the JCT 81 contract provide for? Firstly, as regards precedence of documents neither the Recitals, nor Article 4, nor Appendix 3 is of any help other than the footnote (b) to the Third Recital which confirms that express terms in the contractor's proposals take precedence over different express terms in the employer's requirements, if not resolved prior to contract.

However, JCT Practice Note CD/1B comes partly to the rescue. This provides in the "Commentary on the Form of Contract" in respect of the Third Recital:

"The intention here is that:
1. The Employer accepts that on the face of them, the Contractor's Proposals and the Contract Sum Analysis do in fact correspond with the Employer's Requirements . . . but that . . .
2. this will be without prejudice to the Contractor's liability in respect of the design, selection of the appropriate components and materials and goods and provision of the workmanship to satisfy those Requirements."

JCT Practice Note CD/1B then very helpfully gives an example:

"If the Employer's Requirements include a heating system which was to achieve certain temperatures, the Employer's acceptance of the Contractor's Proposals as including a heating system would not necessarily imply that the Employer accepted that the heating system proposed was adequate to achieve the required temperatures. If therefore it was not so adequate, the Employer would not be precluded by the Third Recital from alleging breach of contract for such inadequacy."

JCT Practice Note CD/1B then continues:

"The Contract Conditions do not deal with the position where, despite the Third Recital and the advice at footnote (b) there is a divergence between the Contractor's Proposals and the Employer's Requirements. It was considered more appropriate to emphasise the need to follow the advice in footnote (b) than to include any specific provision on such divergencies."

Apart from noting that the JCT drafting committee have apparently got a language cross-over in respect of "divergencies" and "discrepancies" I find the above advice, particularly the heating system example, most instructive as to the philosophy of design and build, i.e. the employer stating his requirements in performance/appearance terms to be achieved and the contractor then being required to work out *how* they will be delivered.

Finally in this section it might be helpful to summarise the JCT 81 contractual provisions covering discrepancies *within* documents:

2.4.1 • Discrepancies *within* the employer's requirements *and* identified in the contractor's proposals—contractor's stated assumption to prevail, unless

• post-contract the employer wishes to adopt a different solution, in which case he must issue a Clause 12 change, with the attendant need to agree any alteration to *cost* and *time*.

• Discrepancies *within* the employer's requirements *not* identified until after the contract has been entered into, or arising due to a previously issued Clause 12 change, shall be notified in writing as soon as possible after discovery by the contractor to the employer, together with a proposal for resolution.

• Such post-contract discrepancies shall be resolved at the contractor's discretion unless the employer directs otherwise but in either case it ranks as a Clause 12 change.

2.4.2 • Discrepancies within the contractor's proposals shall similarly be notified in writing as soon as possible after discovery by the contractor to the employer, together with a proposal for resolution.

• Such resolution may be accepted or amended by the employer as a Clause 12 change but at no additional *cost*—and presumably with no implication as regards *time*.

2.4.3 • Equally if it is the employer, rather than the contractor, who spots the discrepancy within either document, the notification requirements are reversed, but the Clause 12 change status remains the same.

8.3 DESIGN RESPONSIBILITY

The previously quoted extract from JCT Practice Note CD/1B and the heating system example merely confirm that responsibility for design remains entirely with the design and build contractor, notwithstanding that the employer or his agent may have given the contractor to understand he has "approved" either the contractor's proposals, which might refer to proprietary systems or materials, or such details as are submitted in the course of design development proposals.

For the avoidance of doubt a disclaimer as regards the meaning of "approved" or "approval" should be spelt out in the employer's requirements of any design and build project.

Only in the most specific situations, e.g. where the employer instructs a Clause 12 change and specifies a particular system or material, which the contractor refuses to endorse, will design responsibility move back to the employer.

Under Clause 4.1.1 the contractor may reasonably object to any instruction of the employer and presumably it would be a condition precedent to any such shift in design responsibility that the contractor should have so objected, or at least conditionally accepted the instruction, before any work was carried out on the instruction.

Design responsibility is thus the "First Commandment" of design and build: "You the Contractor shall design and you, the Contractor, shall build. If it doesn't work, or in any way fails at any time in the next 6 or 12 years (depending on whether the Contract is under seal) you the Contractor, shall put it right."

Only if the employer has (foolishly) given the contractor express written absolution in respect of a specific element of the design will the contractor escape this absolute responsibility.

In the recent landmark case of *George Fischer Holding* v. *Davis Langdon and Everest and Others* (1994) ORB 775 the issue of the employer's agent's duty to comment and/or warn on the contractor's proposals and subsequent design development drawings was addressed by His Honour Judge Hicks.

In the above case it was an express term that the Employer's Agent would at the pre-contract stage:

"1. Carry out an appraisal of the design drawings and documentation available and as provided by the design and build contractor and to highlight any aspects of the design proposals which he considers to be unsatisfactory and which might be seen as presenting potential problems at a later stage."

A fairly typical term of appointment for an employer's agent which was then complemented seemingly by a standard 28 item "shopping list" of post-contract duties, including the duty to certify practical completion.

Specifically listed was the duty to:

"Make visits to the site sufficient to monitor the contractor's workmanship and progress; to check on the use of materials, to check on the works conformity to the specification and drawings and to report generally on the progress and quality of the works having regard to the terms of the contract between the employer and the contractor".

DLE, as employer's agent, relied on the duty to "appraise and highlight" aspects of the contractor's design *as being limited to the pre-contract phase*. The Judge was certainly not impressed, holding DLE liable for failure to identify the patent defects in the alternative roofing system proposed by the contractor—which centred on the length of slope required, the minimal fall as designed, and the need for "end laps", i.e. cross-welts. There was also the question of whether, even if the "end laps" had been correctly formed, this design of roof could ever have been reasonably expected to be watertight given the allowable tolerances in the underlying structure (notably the steelwork), compared with a prefabricated but more expensive "no end laps" system, with each sheet in one clear length.

On the facts, His Honour Judge Hicks found DLE were responsible

" . . . for such matters as approval of substitute materials, consent to deviations, consent to additional drawings and details and instruction of variations equally required approval of design considerations."

Arising from this judgment, there are two important messages:

1. For employers and solicitors representing employers: make sure you lock the employer's agent into such design responsibilities in equivalent express terms.
2. For employer's agents—often quantity surveyors or other professionals whose primary skill is not as designers of buildings: do know when you are out of your depth and bring in, or sub-contract, the necessary expertise—and avoid the very real risk of getting it wrong.

8.4 QUALITY CONTROL

Previous chapters have focused on the need for laying down "ground rules" in the employer's requirements, i.e. standards to be achieved by way of performance specification, or where the *quality* is optional by way of price benchmarking and the use of prime cost sums by the employer.

There is of course nothing to stop the contractor also clarifying the quality he proposes by also stating prime cost sums for the materials involved. It will then be for the employer to satisfy himself that the PC price proposed by the contractor is adequate, and then to select the material within the stated price.

As regards post-contract control of *quality*, i.e. material selection where not specified, workmanship and compliance generally, the employer may well be badly exposed, especially where the design and build contractor has to find savings to make good a deliberately keen tender.

Quite possibly the appointed employer's agent may be suitably qualified himself to judge *quality*, i.e. an architect or building surveyor, but he is unlikely to be site resident except on the largest projects. If only a periodic visitor he will be on site for probably 10 per cent of the time at best. This 90 per cent period of absence is a window of opportunity for the less scrupulous design and build contractor, particularly where it comes to substitution of inferior and cheaper materials. It is therefore important that the employer recognises this risk and ensures that the employer's agent implements effective site supervision.

Notwithstanding the additional fee commitment involved, the employment of a full time, or at least a visiting, clerk of works should prove value for money, provided a no-nonsense ex-tradesman, who can read and write competently, can be found. However, be warned, do not assume all such ex-tradesmen can read and write.

The important thing is that the contractor should not know when he is being watched. It is no good, if as on one project I had, the nice old boy was known never to turn up until mid-morning—and then he would have a cup of tea and a chat! If he had ever turned up at 8.00 a.m., as I occasionally did, he would have been mortified to find how many operatives were not even on site, or at least were not working by the start-time claimed on the time sheets which he regularly signed "As a record"!

If a clerk of works is to be appointed, do be clear as to:

- Who is his employer, i.e. the employer's agent *or* the employer?
- To whom he reports;
- The requirements for specific site records and documentation to be produced by the contractor.

On one particular project, when I was brought in as a claims consultant, I interviewed the clerk of works, only to find that his idea of keeping a diary was making the odd note in his personal pocket diary!

Likewise if the employer's requirements provide for Concrete Cube Tests and Fire Door Certificates do help the clerk of works by separately scheduling out his duties and the specific quality control documents required, which the contractor may or may not volunteer without prompting.

As we have seen at section 3 of this chapter, the employer's agent has a continuing duty of design approval throughout the project, if such a duty can be inferred from the express terms of his appointment—*George Fischer Holding Limited* v. *Davis Langdon and Everest and Others* (1994) ORB 775.

The plight of the employer's agent in that case was compounded by a plea of mitigation along the lines of:

- Even if more site visits had been made, it would not be reasonable to have expected the employer's agent to have picked up whether the critical "end laps" across the roof panels were being installed with absolute precision.
- Where there is a defect waiting to happen (in this case the omission by the contractor of a movement joint around the build-in purlins as shown on the contractor's own drawings) but it has not yet manifested itself (in this case, by cracking of the blockwork) it is not yet a defect.

His Honour Judge Hicks was particularly severe on DLE on both these counts, especially when a witness suggested he could not have inspected the roof as there was no safe access!

So again, all you employer's agents, get out there, climb all over the place and if you have no head for heights, get yourselves a professional clerk of works. Better still, find a clerk of works and get the employer to employ him direct—it might confuse issues of liability nicely if things go wrong.

On the other hand, any employer or solicitor who falls for this direct employment ploy will only have himself to blame if things go wrong. Make the employer's agent responsible for the clerk of works.

8.5 DEFECTS

The down side of lack of quality control is defects—and in the context of design and build defects are not limited to workmanship and materials. There are not infrequently serious design defects—mechanical and electrical services installations being a regular problem area—which can range from compliance issues to function of the building. Needless to say, they are not easily or cheaply corrected once built in!

Under JCT 80 there is often a problem of fault allocation, i.e. is it design or is it workmanship? In real terms this translates into: Whom does the employer sue when there is a major problem? Is it the architect for design, the contractor and subcontractors for faulty installation or the architect again for failure to supervise? In a big dispute such as a fire loss recovery case there may well be 15 or more levels of defendants, with a whole variety of alternative pleadings and Scott Schedules which go on for ever and disappear somewhere to the right on a computer spreadsheet.

With design and build, and JCT 81, all this is avoided—the employer can enjoy *single-point responsibility*. It follows that a dispute as to any alleged defect is as between employer and contractor, and it is *the latter* who then has the problem of fault allocation as between architect, engineer, sub-contractors, etc. Provided the employer can substantiate any alleged defect, there should be no hiding place for the design and build contractor.

The responsible design and build contractor will put his hand up immediately to the employer and put remedial works in hand irrespective of the internal witch-hunt which may be necessary. Unfortunately, experience shows that not all design and build contractors are quick to put up their hand—they tend to play for time while they sort out who is going to put the defect right without cost, or at minimal cost, to themselves.

Such contractors will then advance excuses, arguments, etc. to the employer that either the alleged defect is not a defect, or that it can be patched up, so to speak, rather than properly made good, or that it is somehow attributable to a change in the employer's requirements.

Given this scenario the employer has problems which can take either, or all, of the following forms:

- Is the *quality* or detail required in the contract documentation clear beyond reasonable doubt? Often not, sadly.
- Is the *quality* of finished work less than that required? Often room for opinion.
- Does the alleged defect warrant not granting practical completion? And all that entails.
- How can the alleged defect be effectively remedied and when? What then happens to manufacturers' warranties etc?
- Does the alleged defective work qualify for interim payment under the valuation rules? If so, what value?
- Does the employer have the option of accepting the alleged defective work? If so, can he arbitrarily only allow a discounted value?

Whilst the above questions might equally arise under a JCT 80 contract, the problem under the JCT 81 "soft" version is that there is no designated third-party umpire, charged with seeing fair play between employer and contractor. Human nature being what it is both employer and contractor may have directly opposing views to any or all of these questions given a particular defect alleged by the employer.

Under the "soft" JCT 81, unless such matters can be mutually agreed, there is no alternative to full-scale arbitration, assuming Clause 39 has not been deleted, or litigation if it has.

If, however, the Supplementary Provisions have been incorporated and you therefore have a "hard" version of JCT 81, then at least one has recourse to a third-party umpire if needs be in the form of the adjudicator, *but* only in respect of disputes notified prior to practical completion.

The chances are that practical completion has been achieved and the employer has taken occupation and then defects appear—either visual defects in the perception of the employer or more practical matters such as imbalances in central heating and air conditioning.

JCT 81 Clause 16.2 provides the basic defects sign-off procedure:

- No later than 14 days *after* the expiry of the defects liability period stated in the contract (usually 12 months) in relation to the previously certified date of practical completion the employer must issue "as an instruction of the Employer" i.e. in writing, a Schedule of Defects.
- Such defects shall then be rectified by the contractor within a reasonable time, at no cost to the employer unless . . .
- The employer shall otherwise instruct that remedial works are not required and that he will require "an appropriate reduction . . . from the Contract Sum".

Alternatively, JCT 81 Clause 16.3 empowers the employer to call for the rectification of any alleged defect *before* the expiry of the defects liability period, with the same proviso that the employer may so notify a defect, but accept such defect subject to an appropriate reduction in the final account. A further proviso is that after 14 days from the end of the defects liability period the employer is shut out of either remedy if he fails to serve the required Schedule of Defects on the contractor.

As to whether the contractor then competently carries out the instructed remedial works can be another problem area, particularly where such remedial works involve design elements. Sensibly the employer's requirements should provide that in respect of remedial works the employer may require the contractor to submit method statements, including design proposals, as to how and when remedial works are to be carried out. It should, however, be made clear that any acceptance thereof is without prejudice to the contractor's continuing design liability.

Such a requirement for method statements in respect of remedial works is of course particularly relevant if the extent or nature of such remedial works fall within the ambit of the Construction (Design and Management) Act 1994, otherwise known as the CDM Safety at Work Regulations.

Finally in this section, do remember the "start point" concerning practical completion. If there is a known defect which cannot be simply and conveniently rectified on an agreed visit basis, then practical completion has *not* been achieved—but then life in the construction industry is rarely simple, so please refer back to section 7.3 for my summary of what the current case law suggests after the decision in *George Fischer Holding* v. *Davis Langdon and Everest and Others* (1994) ORB 775.

Above all else, if, as an employer's agent, you find yourself caught in the cross-fire between employer and contractor throwing conflicting contract clauses or interpretations at one another, *do record all relevant conversations and meetings and if a compromise agreement on Defects v. Occupation is necessary, get it in writing, signed by both parties.*

8.6 SET-OFF OPTION

Finally, attention should be paid to JCT 81 Clause 8, Materials, Goods and Workmanship, and the detailed provisions therein, together with the Code of Practice in respect of procedures to be followed if opening-up is required, all as now incorporated in the standard JCT 81 Form of Contract. If these provisions appear complex, further explanation is given in JCT Practice Note CD/1B.

As previously stated, JCT 81 Clauses 16.2 and 16.3 allow the employer to choose whether to have an alleged defect remedied, or whether to live with the defect, but not to pay full value. In practice this is a very grey area and is open to abuse by the employer, as indeed is the whole concept of defects, given that there is no third-party umpire as there is in JCT 80.

Just the same as there are less reputable design and build contractors, so inevitably there are employers who are more difficult to part from their money than others. If an employer is of the more difficult variety, he may well adopt a comparably higher level of quality and compliance acceptance.

However, assuming defects are genuine and/or admitted by the contractor, how does the employer properly and fairly ascertain the set-off value? He can of course start from JCT 81 Clause 30.2B.1.1: "the total value of work properly executed" and contend that the work being defective it is, *per se*, not "properly executed" and consequently does not qualify for payment, i.e. it is of no value.

Here of course we are getting into deep legal water—deeper than the 7' 6" of the infamous swimming pool dispute between Ruxley Electronics and Construction Limited and Mr Forsythe which went all the way to the House of Lords. Incidentally, can anyone tell me whether the disputed 7' 6" was the specified depth of water at the deepest point where Mr Forsythe wished to dive, *or* the specified depth from the top of the pool surround? On such matters of detail do cases turn!

Getting back to our particular defect situation, if notwithstanding the admitted failure to comply with specification the finished work, although defective, partly does the job, it must have some value to the employer. But what if:

- To rectify it now will cost more than the contract value—is the answer *nil* value payable or is the answer *some* value payable, assuming the employer does not insist on immediate replacement?
- It will involve earlier replacement than might have been the case had it been fully compliant, or it will require ongoing maintenance over and above that which would have been required? How can the value payable be equitably assessed?

Suffice it to say these are the sort of questions which do arise and which can be particularly tiresome and expensive to resolve—as Mr Forsythe discovered.

Under the adjudication provisions of the Housing Grants, Construction and Regeneration Act 1996 a dispute can be notified at any time so the existing JCT 81 provisions limiting the power of the Adjudicator to act only up to practical completion will no doubt be amended by official JCT Amendment.

8.7 TYPICAL PROBLEMS

Problem A

Benchmarking of quality—hidden savings.
JCT 81 "soft" contract, i.e. Supplementary Provisions not incorporated.

The contractor's proposals benchmarked the quality of wallpaper by stating "PC £7.50 per roll".

The contractor then offered a sample board of wallpapers and the employer selected. Nothing was said about costs but subsequently the employer discovered that the trade price was £5.25 per roll. Was the employer entitled to the saving, or was the contractor correct in claiming his £7.50 per roll price was a cost cap only and that he was free to purchase within that price bond?

Problem B

Benchmarking of quality—hidden extras.
JCT 81 "hard" contract, i.e. Supplementary Provisions incorporated.
The employer's requirements specified facing bricks and artificial stonework by respective manufacturers, but were silent as regards jointing and pointing details, other than providing that sample panels were required for approval.

The contractor prepared six sample panels showing the combination of facing bricks and stonework, including different colours and details of mortar pointing.

All six sample panels were rejected and six more panels were requested. The contractor then offered a different artificial stone, from another manufacturer, which to all intents and purposes appeared the same. Nothing was said about costs or the change of stone manufacturer and the employer selected sample panel No. 11.

The questions are therefore:

(a) Was this selection of a differently sourced artificial stone and the approval of mortar details a Clause 12 change?
(b) If so, and notwithstanding that Clause S6 had not been complied with, was the contractor entitled in principle to any part of the substantial extra costs claimed on final account?
(c) Assuming such additional costs could be partly justified by reference to invoices etc., could the employer rely on Clause S6.6, and deny additional payment under Clause 12 on the basis that if the contractor had advised the cost differentials at the time of sample panel selection but now claimed on final account he may well have opted for a different, and cheaper, panel?

Problem C

Fitness for purpose (1).
JCT 81 "hard" contract, i.e. Supplementary Provisions incorporated.
Employer's requirements specified windows by reference to an aluminium extrusion manufacturer. Fabricator not specified but windows to be generally double-hung Georgian style vertical sliding casements, with some bottom-

pivoted casements in arched fanlights, all to be capable of external cleaning from within the building.

During the tender period one tenderer sought further instructions as the named extrusion manufacturer no longer produced sections suitable for double-hung casements. All tenderers were then instructed to tender based on the original extrusion manufacturer but using bottom-hung pivot windows, with side hinges operable by key to allow external cleaning from within.

After contract award, one small window was installed by way of a sample and was approved by the employer's agent. A full order of some 150 windows was then fabricated by a local fabricator on a domestic subcontractor basis.

When the main windows were installed, problems emerged, notably the weight of glass in the larger windows, which although within the BS tolerance for size of aluminium framing, caused the window to flex when opened. Such flexing in turn caused the mitred aluminium joints to open up and the head of the windows to hit the blinds and tracking, damaging the blinds and the plasterwork soffit. A further problem was that there was no way of holding the bottom-hung pivot window in the open position—so wind pressures would then slam the windows shut, only to then suck them open again with a severe jolt on the hinge and exaggerating the flexing, before repeating the process until manually closed and latched.

All the windows were therefore condemned as not fit for purpose by the employer's agent, supported by an independent industry expert, and the contractor was instructed to replace all the windows. The contractor refused on the basis that the sample had specifically been approved by the employer's agent and if all were to be replaced it would be a Clause 12 change, costing some £200,000!

As there were insufficient funds in the retention fund and an ineffective performance bond, the employer could not:

- Afford to instruct a third-party contractor to replace all the windows;
- Risk doing so and then having to sue to get the costs back from the original contractor.

So the windows remain and it is just a question of time before they self-destruct and one falls out—which could have all sorts of consequences, including personal injury or worse.

Problem D

Fitness for purpose (2).
JCT 81 "hard" contract, i.e. Supplementary Provisions incorporated.
The employer's requirements called for a "Goods Inward" delivery yard and loading bays laid to falls to suit delivery lorries supplying bulk foodstuffs from manufacturers to the central distribution warehouse, which was the subject of the contract.

The contractor visited a similar facility and then designed and built the service yard based on flat-bed Type IV delivery lorries, such that when the rear lorry doors were opened and the trolley straps released, the laden trolleys would gently roll into the warehouse, where forklifts would take over.

Unfortunately a new Type V lorry was just being introduced by some manufacturers with an in-built slope and it so happened that the very first bulk delivery lorry to roll up at the brand new distribution centre, just 11 months after the cows had been driven from their pasture, was one of the new Type V lorries—full of Bounce dog food.

All the national distributor's top management were there to mark the occasion, attired in crisp white uniforms. A brief speech and the lorry doors were opened, then the trolley straps were released. In a matter of seconds several tons of Bounce dog food was on its way and gathering momentum due to the double-slope effect—scattering the management in all directions before slamming into and severely damaging a concrete encased steel column. The resultant mess and smell as many of the tins burst open, splattering Bounce in all directions, was indescribable.

An urgent instruction was issued to flatten some of the bays in the service yard, which meant a drainage redesign—but who would pick up the cost tab? As the project had been a great success in all other respects a 50–50 cost share deal was agreed in the time-honoured fashion over a good lunch a few weeks later, and the final account was closed after the remedial works had been carried out.

Problem E

Relying on third-party design information.
International design and build project for the Iranian Navy in the mid-1970s—before the demise of the Shah.

The UK contractor was required to ensure the old timber jetty was sufficiently strong to receive the crane off-loading Chieftain Tanks, and did so relying on information from the Iranian Navy.

When the big day arrived complete with brass bands and top navy brass, the first Chieftain Tank was swung ashore onto the jetty. Half-way through the first speech an ominous noise started and the assembled party just had time to get onto terra firma before the whole jetty and one Chieftain Tank slowly toppled into the waters of the Arabian Gulf.

In those days it was not so much a question of who would pay, but who would get shot! But worse was to follow when the formalities were re-arranged two days later at the modern roll-on, roll-off dockside. After the speeches and handover of ceremonial keys the trusted No. 1 Iranian tank driver saluted but then selected reverse gear by mistake, and promptly disappeared over the edge of the quayside! At that stage the score was two down and 28 to go, and the assembled UK expatriates didn't know whether to laugh or cry!

CHAPTER 9

DISPUTES

9.1 PROCEDURES AND OPTIONS

Even on the best-run projects disputes can arise. As we have seen, due to the particular allocation of risks and responsibilities, design and build contracts offer possibly more scope for disputes than most other forms, notably in respect of *quality* and *cost* issues.

As regards time issues, as the design and build contractor has control of both the design information flow *and* the site activities he should really not have too much opportunity to make extension of time or loss and expense claims—particularly if the employer has opted for the "hard" version of JCT 81 including the Supplementary Provisions. Equally where under the "soft" version the employer's agent retains firm control by requiring monthly progress reports, including contractor's completion programmes, the design and build contractor should only have limited opportunity for making valid claims on the employer.

We have seen how the "hard" version of JCT 81 has its own in-built disputes procedure by way of adjudication at Supplementary Provision S1, but what if the employer has not opted for the Supplementary Provisions?

If Clause 39 in the standard JCT 81 contract has not been deleted then the formal disputes procedure is arbitration. Alternatively, if Clause 39 is deleted the parties are free to litigate but still might be referred back to arbitration following the decision in *Northern Regional Health Authority* v. *Derek Crouch Construction*.

However, with the absence of a third-party umpire under JCT 81 it can be very costly even for the "winning" party to have to go through arbitration or litigation to resolve what might be a simple technical dispute, e.g.:

- Is the work properly executed—so that it should therefore be paid for?
- Is the alleged instruction a Clause 12 change or not?
- If so, how should the change be properly valued?

The question is therefore: how can the parties most effectively resolve their differences *and* retain a good working relationship? Whilst the contract might provide for arbitration or litigation there is nothing to stop the parties mutually agreeing one of various alternative dispute procedures, so limiting their cost exposure in engaging solicitors, experts and counsel.

Perhaps more importantly they can obtain immediate access to an alternative procedure, if they so agree, but at the same time agree in principle whether or not the decision to be obtained by such alternative dispute resolution procedure is, or is not, to be binding.

Sensibly, the most appropriate stance is that any such decision is binding, but only up to practical completion. If the "losing" party still feels aggrieved after due passage of time, it is then open to him to institute arbitration or litigation proceedings—and to have to fund up front the very substantial costs involved, with the attendant risk that his action might not succeed. In that event he will be looking at having to pay a very substantial part of the other side's costs, as well as his own.

As such, arbitration or litigation is not for the faint-hearted or the impecunious—but therein lies one of the great problems of the construction industry: how does a subcontractor who has been forced close to, or actually into, insolvency by a main contractor not paying sums arguably due find the further money required to enforce his contractual remedies?

Obviously on any project there may well be genuine areas for disagreement, but the main contractor who pays 75 per cent of subcontractor's interim payment applications, often without a detailed statement as to monies disallowed, and then virtually cuts off all payment as practical completion approaches, is a commercial thief, and should be criminally prosecuted. Unfortunately even some of the better known names in our industry, whilst professing to be "best practice" contractors and advocates of "partnering", are guilty of such dishonest practice.

Hopefully a greater awareness of dispute resolution procedures, notably adjudication, will help to end such malpractices, and with developments in information technology I personally see no reason why the whole interim valuation and monthly payment procedure should not be totally transparent right down the contractual chain.

However, recognised and proven alternative despite resolution (ADR) procedures can be listed as follows:

- Adjudication
- Expert determination
- Mediation
- Conciliation

Adjudication is discussed in detail later in this chapter, and the other three are really variants on the same theme, i.e. a respected intermediary who can be called in very quickly. Obviously both sides must know and observe whatever ground rules are put in place, and show good faith if a result is to be achieved. They must have a genuine desire to settle their differences and not just go in on a "Heads I win, tails you lose" basis, i.e. if they do not come out substantially on the right side they abandon the process and revert to full arbitration or litigation.

In this respect, the key to success of such alternative dispute resolution procedures is that the parties are represented not by the individuals who have personally fallen out at site level, but by a higher level of management who, with the benefit of some distance, can more readily take a wider commercial view.

Trained individuals who can act as experts, mediators or conciliators are available by reference to such bodies as CEDR, the Centre for Dispute Resolution, or the RICS, the Royal Institution of Chartered Surveyors, to mention just two such bodies.

9.2 ADJUDICATION

Adjudication has been a recognised form of first-stage dispute resolution procedure for over 25 years in the construction industry, having been introduced by the JCT in respect of disputes under the old Green and Blue subcontract forms, but only in respect of set-off, i.e. the right of main contractors to withhold payments from nominated subcontractors and nominated suppliers.

Over the last 15 years employers and architects have preferred to pass all risks of non-performance by favoured specialist subcontractors on to main contractors, i.e. most subcontracts, however specialist, are now placed on a domestic subcontract basis. As such nomination is now rarely used and in consequence adjudication has virtually disappeared.

The introduction of the Supplementary Provisions into JCT 81, by virtue of Amendment 3, February 1988, with much wider powers for the adjudicator under clauses S1 and S7, has led the revival—and adjudication has now been adopted by the Housing Grants, Construction and Regeneration Act 1996, following the Latham Report.

These are early days, with this new Act due to come into force in May 1998, so details of how adjudication will work in practice remain to be seen, but in the fullness of time there will hopefully be a consensus document, e.g. ORSA Adjudication Rules, which will provide for:

- How adjudicators are selected and appointed;
- How adjudicators are to act and what if the dispute is not simple;
- What happens if the appointed adjudicator cannot reach a decision within the required timescale;
- How the "winning" party can enforce the adjudicator's award;
- Whether the adjudicator's award is subject to appeal by way of arbitration or litigation when the project reaches practical completion;
- How the adjudicator recovers his fee.

In my opinion, the present provisions of the Act fail in two respects. Adjudication is only an *optional* first-stage dispute resolution procedure and that being so, one party can save up his grievance, preparing a very detailed and fully documented claim which he then launches many months later on the other side and calls for adjudication.

The other side is therefore at some disadvantage given the time constraint —usually 28 days for the total procedure—in that they have to prepare their reply in detail *and* submit same to the adjudicator, who *then* requires time for consideration and investigation etc. before publishing his decision, all within the said 28 days. The adjudicator is therefore likely to be faced with a great imbalance of evidence and detailed submission.

I would prefer to see adjudication made a *compulsory* first step, before either party can call for arbitration or litigation. I would also like to see a provision that any event or issue cannot be referred to adjudication more than three months after the originating event. In this way deliberate "ambushes" will be almost impossible.

As to the details of the adjudication procedure, time is of the essence and I believe best practice will require solicitors setting up contracts on behalf of employers and funders to avoid the potential delay involved in seeking institutional appointments by naming adjudicators in the head contract. As disputes may cover anything from *cost* to *quality* to *time*, as well as design issues and professional negligence, there is an argument for naming, say, three adjudicators, who cover the range of required skills. Such a triumvirate could then be either called upon individually by the parties, or called upon to act as a committee.

However, the most important requirement in my view in respect of any adjudication provisions written into head contracts is that such provisions are passed down through the contractual chain, i.e. "single-point adjudication", meaning that the same names and ground rules *must* appear in all dependent subcontracts. If not, the main contractor will be able to rewrite the names and the rules for subcontract purposes, and any dispute which affects more than two parties may well be adjudicated with conflicting and incomparable results —"win-win" and "lose-lose" scenarios being a distinct possibility—which would be a manifest injustice and could well discredit the whole adjudication process.

9.3   ARBITRATION

The Clause 39 provisions of JCT 81 require that arbitration proceedings can only be commenced once practical completion has been certified, except:

- Where practical completion is itself disputed; or
- Where the works have been abandoned or determined.

Assuming therefore that the above scenarios do not apply and that there is a typical dispute concerning validity of a Clause 12 change, and its effect if valid on *cost* and *time*, Clause 39.2 makes a further exception, recognising the need for immediate arbitration, in the absence of a third-party umpire.

In this situation, immediate reference may be made by either party where the issue is any one, or a combination, of the following:

- whether or not the issue of an instruction is empowered by the Conditions, or
- whether or not a payment has been improperly withheld, or
- whether a determination under Clause 22C4.3.1 will be just and equitable, or
- whether either party has withheld or delayed a consent or statement where such consent or statement is not to be reasonably withheld or delayed.

Further exceptions are provided for under Clauses 4.1.1, 8.4, 17.1, 18.2.2 and 23.3.2 concerning disputed reasonable objection to required consents, i.e. immediate reference to arbitration is allowed in these events.

However, the reality of an arbitration reference is that it is not immediate. Firstly, the parties will inevitably try other avenues and be reluctant to fire the first shot in what could be an expensive and losing war. Secondly, the dispute in the meantime usually escalates in quantum and festers in terms of site relationships.

Only in desperation will one party serve Notice of Arbitration—usually the party who perceives he is being kept out of money—and by the reverse of the coin the party who arguably has retained the money does not want to be parted from it. Accordingly it is not usually in the respondent's interest to co-operate in the arbitration process, so typically the respondent will obstruct the claimant by:

- Taking his time in *not* agreeing the list of potential arbitrators put up by the claimant;
- Then propose his own list of potential arbitrators, choosing individuals who are unlikely to be acceptable to the claimant;
- Then introducing a deliberate red herring of maybe trying one of the various forms of alternative dispute resolution previously mentioned——with no intention of actually adopting such a course of action, but with every intention of persuading the claimant to further delay his application to the institutional body named as the appointor in the contract;
- Even when, say one month later, an arbitrator is appointed, in not replying promptly to letters agreeing a suitable date for a preliminary meeting, and then being difficult as to the date and place of such a meeting.

In the meantime the dispute is now nearly three months old at least and gathering momentum like a forest fire. Further, if the project is still in the build stage valuable resources which should be focused on completing the job and agreeing the account are diverted to claim elaborations or claim defence tactics. Solicitors and experts are then required, if not already on the scene, and the bills for professional fees start rolling. All too soon the quantum of the original dispute can lose all proportion to the legal costs already incurred in professional fees, plus the costs yet to be incurred in continuing the arbitration up to and including the hearing.

Even before the Arbitration Act 1996 there were a few robust arbitrators who when first approached would insist that in the interests of both parties there should be a short timescale for case preparation, and then by Orders for Directions laid down a relatively short procedure and hearing duration. However, it has to be said that most arbitrators saw themselves as quasi-judges and therefore stuck with the full procedures of case preparation and trial, which incurred considerable time and costs for all concerned.

Happily the Arbitration Act 1996 has taken a radical approach and has now empowered arbitrators to be more robust—subject of course to the necessity of even-handedness and courtesy to the parties. One of the most powerful new provisions is the power for arbitrators to cap costs, i.e. arbitrators can determine the level of costs that either side may spend and accordingly may expect to recover, subject to court taxation, if they are successful. This however does not stop either side spending much more, but of course they cannot recover such further costs if they win.

It is of course early days for the Arbitration Act 1996 and we can but wait to see whether the new rules actually achieve a distinctly quicker and cheaper alternative to litigation, and hopefully a consistently higher quality in arbitrators appointed by the various professional bodies. Above all else, justice must be available to the claimant who has say a claim between £50,000 and £100,000, but who simply cannot risk £50,000 in fees pursuing his arguable entitlement—and more if he loses. Also, more use ought to be made of litigation insurance—both pre-event, which is not expensive, and post-event, which, although relatively expensive, is a defined and limited risk.

Equally, another abuse when the respondent applies to the arbitrator (under the 1996 Act) for security for costs to be put up by claimant, whose money he has arguably withheld for many months, yet when the arbitrator finds for the claimant, the respondent simply announces he is regrettably trading insolvently and calls in the receivers. In these circumstances the claimant has suffered a double injustice.

I would therefore suggest that for construction cases the legal precedent should be reviewed, i.e. if a claimant can persuade the arbitrator that there is reason to believe that the respondent might not meet any award which might be given against him, that the arbitrator should consider ordering the respondent to place all or part of the disputed money in a separate trust fund—with the respondent having the right to take the interest thereon in the meantime.

This might be contrary to current court practice but in this modern age, with widespread payment abuses of subcontractors by main contractors in the construction industry, I ask—why should *only* the respondent, or defendant, be able to apply for security for costs?

How often in the construction industry today can David afford the sling and stone with which to bring Goliath to his knees? At least under the Arbitration Act 1996 arbitrators can hear and decide such applications for security for costs.

9.4 LITIGATION

In the scale of things litigation, i.e. High Court action heard in the Official Referee's Courts, should be reserved for the bigger value disputes, or for where a matter of serious public concern is at issue, e.g. a significant point of law or an allegation of dishonest conduct.

Just as the arbitration process has been overhauled by the Arbitration Act 1996, so the High Court has not been insensitive to considerable criticism in respect to delays, lengths of trials and uncontrolled costs. The construction industry is of course very lucky to have its very own branch of the High Court where most of the judges have come up through the ranks of having been specialist junior and then senior counsel, dealing almost exclusively in construction disputes over many years.

As such the quality of the procedure and eventual judgment is second to none, but it comes at a heavy price in terms of wasted time and costs directly and indirectly thrown away. Such is the backlog of listed cases in the London Official Referee's Court that even to get a listing for a short four-day trial may well involve a two-year waiting list, or being listed sooner but as a second or third fixture.

Almost inevitably when the big day comes one key individual is unavailable or something has happened, and an adjournment has to be sought—usually one of the counsel is double-booked in another ongoing action which should have settled but has not! In a typical multi-party case there may well be a plaintiff and three levels of defendants, each with senior and junior counsel, i.e. eight barristers who all have spent considerable time and money learning their parts and who all need to be available at the same time—not to mention key witnesses.

Court lists are therefore somewhere between a juggling act and a lottery, and it is this delay and uncertainty as to when the trial will actually begin which creates so much wasted effort and unnecessary costs for both parties, e.g. refreshers paid to counsel, experts' reports whether written two or more years previously and which now have to be dusted down and revisited. In my opinion this is the one major downside to litigation, and but for this factor I believe many more arbitration clauses in standard contracts would be struck out in favour of litigation.

The High Court is therefore addressing this major criticism by virtue of the Woolf Report which has radical proposals for streamlining the procedure by way of:

- Proactive pre-trial case management by the Official Referees, e.g. identifying and hearing preliminary issues;
- Restricting "discovery" to "standard disclosure" only, unless further disclosure is ordered;
- The use of experts jointly instructed by the parties;
- Focusing on the need for impartiality of expert evidence;
- Giving both the plaintiff as well as the defendant the right to make a sealed offer backed by a sliding scale of interest rates related to value if the offer is not accepted.

Hopefully the above will be complemented by the appointment of more Official Referees, maybe by persuading retired judges to work part-time in dealing with pre-trial matters.

However, on a personal note, I am concerned that it might be dangerous to limit the discovery process as, typically, in a construction case there is a big imbalance of paperwork held by the respective parties in the early stages. In my experience it is usually the plaintiff or claimant who has the much smaller pile *and* who has the fewer skeletons in the cupboard. Often only by obtaining full discovery from the defendant or respondent can the plaintiff's or claimant's pleadings be substantiated, and the vital evidence—so long suspected but closely guarded—be flushed out.

Certainly discovery must be controlled in terms of time allowed for inspection, but to restrict discovery to "standard disclosure" only would, I suggest, be likely to prejudice plaintiffs far more than defendants—and such an imbalance cannot be in the public interest.

Whilst the foregoing comments have been of a general nature across all forms of contract dispute, they are particularly applicable to design and build problem projects—where, inevitably, virtually all the evidence is held by the contractor. As such I trust they may be helpful in focusing on the risks and tactics to be adopted, when the normal process of negotiation, with or without the Supplementary Provisions of adjudication, has failed.

9.5   TYPICAL PROBLEMS

By its very nature a construction contract, other than in its simplest, domestic form, is a complex jigsaw of interlocking contracts where each participant may only be in a contractual relationship with his immediate neighbour on either side, but where, nevertheless his own ability to perform is dependent upon the good behaviour of other neighbours further afield.

This dependence may well be in terms of physical reliance on a preceding trade activity, without which one cannot begin to perform one's own obligation, e.g. supply and fixing of profiled sheet-steel decking as permanent shuttering for *in situ* floor slabs. But this activity must in turn depend upon the prior erection, checking and handover of the structural steel frame for the relevant area of work.

Problem A

A dispute had arisen concerning the proper final account evaluation and loss and expense claim for the supplying and fixing of profiled sheet-steel decking which, instead of taking 19 weeks, took 52 weeks, with attendant loss of productivity etc.

Subcontractor C (profiled decking) therefore served Notice of Arbitration on subcontractor B (structural steelwork), who in turn was already in a much wider ranging dispute and arbitration with the contractor A. Under the terms of the subcontract B passed up C's claim to A, repeating all allegations and figures verbatim, but failed to formally deny C's claim. C went to the High Court for summary judgment, but the dispute between B and C was referred to arbitration.

However, as arbitration is private, C was unable to find out where or when the arbitration between A and B was taking place. Even if C had found out, the arbitrator in the dispute between A and B would probably have not allowed him to attend, even as a silent spectator at the back of the hearing. How therefore could C know whether, if at all, and how strongly B had pressed C's claim on A? Further, what if B's claim in respect of late erection of structural steelwork due to A's alleged failure concerning design information failed—it must follow that A could not be held responsible for C's alleged losses and accordingly C's case may never have been properly addressed in the arbitration?

In the meantime C was paying bank interest on his losses, was paying ongoing professional fees, and generally his good case was going nowhere fast. B had total control and could simply keep fobbing C off on the basis, rightly or wrongly, that the arbitration with A was ongoing. Can this be justice at work?

Alternatively, could C start a parallel arbitration against B with a different arbitrator? If so, could B have reasonably objected on the basis of either:

(a) Unreasonableness of being required to handle two parallel actions?
(b) The risk of conflicting decisions, i.e. lose/lose?

In the above circumstances I believe there needs to be some mechanism for giving C a statutory right to know all relevant details of the upstream arbitration, and to be represented if he so wishes.

Problem B

In most claims for disruption or loss and expense it is usually the timely issue of drawn information that is the central issue. Consider therefore the following:

1. A major design and build project.
2. The contractor subcontracted the mechanical and electrical design to an outside consultancy. Work started on site, and then a major dispute occurred between the contractor and the outside design consultancy, who then resigned or was fired.
3. The contractor then appointed another outside mechanical and electrical consultancy.
4. The project proceeded in a totally *ad hoc* fashion with various mechanical and electrical subcontractors working hand-to-mouth due to lack of drawn information and consequent inability to resource materials. Much work was abortive and major claims for disruption and loss and expense ensued.
5. The contractor and the second outside design consultancy then had a major bust-up, the latter seeking payment for their unpaid original design fees, plus additional design fees, and with the former counter-claiming for alleged late and inadequate design.
6. The subcontractors discovered this fact and suspected the contractor was running contrary arguments, but what could they do to prove it and obtain a quick commercial settlement?

Effectively the contractor was "running with the hare and hunting with the hounds"—contending no delay or fault on his part with the various mechanical and electrical subcontractors, yet pleading the exact opposite with the outside design consultancy, and contending they were the cause of the delay. Could he legitimately do this? Yes, he had done it, but how within the due process of contract law could he be brought to heel before all the variously affected subcontractors had to give up or be forced to risk full litigation or arbitration? Alternatively what if some of the subcontractors had already had no option but to call in the receiver? How can a receiver, who by definition has limited funds, secure justice for the secured and unsecured creditors?

If writs had been served in the action between the contractor and the outside design consultants those would have been in the public domain. Otherwise it would be left to the aggrieved consultant and one or more of the aggrieved subcontractors to exchange information and full pleadings, thereby boxing in the contractor. Even then matters could have been further complicated by one action being in the High Court and one being in arbitration.

There is therefore a clear need for effective joinder provisions, which hopefully will deliver cost-effective justice and decisions well inside three years, which is about par for the course at present.

At present arbitration may well be quicker than litigation, especially in the time taken to get to trial, but it cannot in practice readily handle multi-party actions, so it is a case of Hobson's Choice. Thus in multi-party actions the High Court, i.e. the Official Referee's Court, is the only effective choice, albeit the slower route—and that is assuming there is no binding arbitration clause in the contract prohibiting such choice.

CHAPTER 10

COMPARATIVE CONTRACTS

10.1 GUARANTEED MAXIMUM PRICE VARIANT TO THE JCT 81 FORM OF CONTRACT

Given the option as previously described of a "hard" form of JCT 81 With Contractor's Design, incorporating the Supplementary Provisions and Clauses S1 to S7, one might question why there is a need for an unofficial Guaranteed Maximum Price (GMP) variant of design and build.

In my view all that is necessary is for the employer to set a realistic level of contingency expenditure and not to stint on the amounts stated in the employer's requirements for provisional sums. Thereby he sets out his stall as to what his overall funding constraints will allow prior to tender. If tenders then come in as expected, or lower, well and good. If they come in higher, the employer has a further option to revist his budget and to review the required scope of the project prior to entering into contract.

From that point in time the twin goalposts of *cost* and *time* are fixed, and by virtue of the provisions of S6 and S7 these twin goalposts can only be moved with the express written consent of the employer. Why then the need for a GMP variant of the officially recognised JCT 81 With Contractor's Design Form of Contract?

There is a school of thought which believes that to ask any contractor to give a fixed price lump sum tender before the development of schematic design is too risky for *both* employer and contractor, i.e. on the one hand it invites low price gambles by contractors, which usually rebound on the employer, or on the other hand it invites safe cover prices by contractors, which represent premium prices for the employer. Potentially if either tactic is adopted by the contractor at time of tender the employer is likely to come second best. Thus there is a perceived need for the employer to *retain budget control in parallel with design development,* and in so doing allow the contractor to price the work more accurately.

Accordingly the concept of a GMP design and build contract has evolved, and such a contractual arrangement is in reality a first cousin of management contracting, in that:

- The employer procures the design and build project based on a first-phase schematic design and performance specification as set out in the employer's requirements—as for the standard JCT 81; but

131

- The contract sum analysis has various elements covered by prime cost sums, which are then tendered as a second-phase exercise, or it can be wholly made up with prime cost sums, in which case the contract sum analysis is the equivalent of the estimate of prime cost (EPC) under the JCT 87 Form of Management Contracting.

The contractors, as part of their tenders, are required to signify their acceptance of the reasonableness of such prime cost sums as against the employer's performance specification. They are also required to agree to go to open competitive tender, vetted by the employer, as and when they have sufficiently developed the design to allow fixed price competitive tenders in those specific work elements.

In this way the employer does not dirty his hands with any design responsibilities—the first principle of design and build—but he also obtains a tendered "best buy" in each element of the works.

In return the contractor receives a pre-agreed mark-up on the finally accepted elemental price *and* a final result bonus according to a pre-agreed ratio of overall project costs saved, e.g. 75 per cent employer: 25 per cent contractor, in the event of the final account coming in under the project-agreed GMP.

On the other hand if the GMP is exceeded the employer pays no more, leaving the contractor to explain why the agreed cost cap has been exceeded. Only in exceptional circumstances will a retrospective claim for loss and expense under a GMP contract ever see the light of day, as it should be clear long before the completion of the prime cost validation process as to whether the agreed contract sum will be achievable, and if not, why not.

Assuming the procurement process is completed and the overall budget is under control it will only be wholly exceptional site events, or poor management by the contractor, which will let in any basis of a loss and expense claim from elemental works package subcontractors. It will then be for the design and build contractor to convince the employer that the underlying cause was exceptional site events as provided for under the contract, rather than the contractor's poor management of the design process.

Although JCT 81 is the most commonly used standard form of design and build contract applicable to the general construction industry, there are other standard forms of design and build contract as issued by more specialist organisations such as the Institute of Chemical Engineers, the IChemE, etc.

Whilst therefore there may be different emphasis in allocation of risks and certification procedures there is a common objective: delivery on *time*, for a budgeted *cost* of a fully compliant building, process plant or structure, which the employer is pleased to have commissioned, i.e. acceptable or better *quality*.

I shall therefore now offer some general comment on these other standard forms of design and build contract, by way of focusing on the six key areas as previously discussed in relation to the standard JCT 81 With Contractor's Design Form of Contract, i.e.:

- Benchmarking of *quality*
- Fitness for purpose
- Contract base-line
- Changes, i.e. *cost*
- Delays, i.e. *time*
- Certification

Finally, I will review each alternative standard form of design and build contract in respect of the disputes procedures provided therein.

10.2 THE BPF/ACA FORM OF BUILDING AGREEMENT 1984 (REVISED 1990)

General comments

This "no frills" design and build contract was the joint production of the British Property Federation (BPF) and the Association of Consultant Architects (ACA). Clearly the basic thinking and the structural layout owes a great deal to the more widely used JCT family of contracts, particularly the JCT 81 With Contractor's Design Standard Form of Contract.

Seemingly the BPF and ACA have focused on the key requirements for a design and build contract and have deliberately selected and developed what might be regarded as best practice, including tackling head-on some of the JCT 81 "grey areas", such as "fitness for purpose". Above all else, the clauses are reasonably short, and generally written in plain English.

The Recitals are given by alphabetic reference, where:

A—Provides for a short description of the works;

B—Provides for the statement of the agreed contract sum, before VAT;

C—Provides two core documents, Certain Drawings and a Time Schedule, plus further options, by way of defining the contract documents;

D—Provides that the contract documents shall be part of the Agreement;

E—Provides for the naming by the employer of a client's representative, for the contractor's right to notify reasonable objection within five days of the employer's notice and for the named client's representative's powers of delegation;

F—Provides for the contractor providing " . . . all further drawings, details, documents and information for the execution of the Works . . . ";

G—Defines a "working day", the unit of time referred to throughout the contract;

H—Provides for dividing the works into sections with different dates for completion.

Thus, under Recital F responsibility for design passes to the contractor, like a baton in a relay race, upon the employer signing off the contractor's proposals, Certain Drawings and the Time Schedule at contract signature.

Thereafter the project is administered by the client's representative, but there is no further reference within the standard contract or guidance notes to clarify whether such a client's representative acts as the employer's agent, as he does under JCT 81 With Contractor's Design, or whether he acts as independent certifier of *cost*, *time* and *quality*, as he does under JCT 80 Lump Sum Contracts.

In the absence of such official guidance, which would have been helpful, I suggest that for the avoidance of doubt, the role of the client's representative should be further defined in the tender documents—and that his position should be that of the architect under JCT 80, i.e. independent certifier.

Benchmarking of quality

Among the list of optional contract documents set out at Recital C are the specification and/or the contract bills.

It follows therefore that it is up to the employer to secure, by full description or reference to national standards, the *quality* of materials and workmanship required in key areas, albeit the specification should be a performance specification.

Likewise where it comes to optional selection of *quality*, the required *cost* level of materials such as wall, floor and ceiling finishes should be stated by Prime Cost (PC) price, as for the recommended benchmarking of optional material *quality* under JCT 81.

Fitness for purpose

No nonsense here: Clause 1.2 provides that:

"Without prejudice to any express or implied warranties or conditions, the Contractor shall exercise in the performance of his obligations under this Agreement all the skill, care and diligence to be expected of a properly qualified and competent contractor experienced in carrying out work of a similar scope, nature and size to the Works."

Quite how a contractor can be deemed to be "properly qualified" might be an interesting test case in due course, but I applaud the sentiment in terms of the building element, if not the design element, of the contractor's obligations.

A major "grey area" in the contract provision of the ACA Form would appear to be design capability, unless this is intended to be caught by the aforementioned "properly qualified". Again it might be helpful if this important element were expressly addressed in the standard contract. Not being so addressed, it is an area where the employer may well need to make express and specific provision in the contract documentation. Even then the contractor's resource level and capability, either in-house or via an outside consultancy, needs to be checked out and confirmed as acceptable to the employer before appointment of the successful design and build contractor.

Clause 3.1(a) then spells out the contractor's total responsibility for all drawings, details and documents and Clause 3.1(b) provides:

"The Contractor warrants that . . . those parts of the works to be designed by the Contractor will be fit for the purposes for which they are required."

Full marks to the ACA and the BPF, but such provision does impliedly require that the employer has actually stated the purposes of the building in his tender enquiry—and that he will not change his mind . . . !

Contract base-line

The ACA Form of Building Agreement envisages the employer going out to tender based on:

- Certain drawings, and
- A time schedule

which, when tenders have been received and agreed, become:

- The contract drawings and
- The time schedule to be stated in the signed contract by way of Alternative 1 (single completion date), or Alternative 2 (phased completion).

The time schedule is required to state:

1. Possession of the site by the contractor (Clause 11.1): Date ....;
2. Taking-over (i.e. completion) of the works by the employer and commencement of the maintenance period (Clause 11.1): Date ....;
3. Rate of liquidated and ascertained damages (Clause 11.3) at £.......... per week;
4. Maintenance period (Clause 12.2): Period ....

Presumably at tender stage the employer will need to set out a basic shopping list, i.e. an outline Schedule of Activities covering the main elements of the required work, supported by schematic drawings, which will in due course become the priced Schedule of Activities (see Clause 17.1(a) with reference to valuation of the client's representative's instructions).

This then leaves the question of the contract bills. Given that the whole essence of design and build is that a project should be procured on a performance specification basis, leaving each contractor to use his imagination in respect of design and buildability, the employer cannot produce such a document.

Accordingly, any bill of quantities can only be a contractor's document, submitted in support of his tender, or it can be a second-stage document produced by mutual agreement between employer and contractor. Whether one adopts the priced Schedule of Activities or contract bills, this equates to the contract sum analysis under the JCT 81 With Contractor's Design Form of Contract, whose purpose is essentially threefold:

- As a basis for tender evaluation and contractor selection, with particular reference to checking whether the required scope and *quality* of the works have been properly allowed for by each tenderer;
- As a base-line for agreeing the *cost* effect of variations, i.e. clients representative's instructions, and therefore the eventual final account;
- As a mechanism for assessing interim payments.

In the absence of any suggested standard format for the priced Schedule of Activities, or method of measurement for the contract bills, the employer will need to specify his requirements in the tender documentation.

In practical terms I would suggest tenders under the ACA Form of Contract, BPF Edition, should be required to submit a priced Schedule of Activities with their tender in a prescribed form, e.g. BCIS Elements for Design and Build (see Appendix A) along with the requirement that back-up details will be provided upon request. Alternatively, if tenders are to be obtained by way of contract bills I would suggest they are required to be prepared according to the RICS Principles of Measurement (International) for Works of Construction, i.e. a recognised short form of measurement equivalent to what are generally accepted as "builder's quantities".

Changes

The ACA Form provides for changes at Clauses 8 and 17, Issue and Valuation of Client's Representative's Instructions. Clause 8 gives the client's representative power to "issue" instructions covering a detailed range of matters, but Clause 8 fails to prescribe whether such "issue" shall be in writing.

However, reference to Sub-clause 8.3 specifically provides for the power to issue oral instructions in an emergency—so by implication "issue" must be in writing. This is subsequently confirmed by Clause 23, Notices and Interpretation, and maybe it would be helpful if this general provision could be moved to the front of the contract.

Not to be confused with Clause 8, Clients' Representative's Instructions, Clause 2.2 covers the general obligation of the contractor to produce design development drawings to the client's representative for approval, geared to the time schedule, and Clause 2.3 provides the detailed procedure for submission and return—10 days being the standard turn-around period unless otherwise agreed. This clause needs some reading and in my view does leave the door open to potential disputes in that:

"The Client's Representative shall . . . return one copy . . . to the Contractor together with his comments on it or endorsed with no comment . . . "

So far so good, with the interesting and positive concept of the client's representative having to write "No Comment" and presumably sign and date each drawing as a matter of record. However, Clause 2.3 then continues:

" . . . provided that the Client's Representative shall not comment adversely on any such drawing, detail, document or information which complies with the Statutory Requirements and with the standards of workmanship and materials specified in or to be reasonably inferred from the Contract Documents."

Some will say this is eminently good sense but I would say it would be better not to give such a hostage to fortune. In my opinion it would be better if the contract provision stopped where I have split the text. I can think of one particular design and build contractor who would trade on this clause and challenge the client's representative's opinion, at every opportunity, thereby seeking to promote an inflated final account, and inevitably a delay claim.

So, imagine you are the unfortunate client's representative fending off such a contractor. What are your powers? Unfortunately the contract is silent, other than at Clause 2.4 which says the contractor has no option but to resubmit his proposals in the light of your comments—but I, as the awkward contractor, have already challenged your original decision as being outside the provisions of Clause 2.3, so Clause 2.4 does not apply, does it? In other words, it is an impasse, and confrontation beckons.

I will deal with the adjudication and arbitration provisions of the ACA Form in due course under "Disputes", but suffice it to say in respect of the above impasse, I do not see that adjudication is necessarily an available solution, Clause 25.2(c) being close, but arguably not close enough. If that is right, then arbitration under Clause 25.6 is also not an option, i.e. the awkward contractor and his solicitor will say Clause 25(c) relates to quality of work being built, not to the validity or otherwise of the client's representative's decision on the contractor's design development proposals under Clause 2.3! I therefore say again, the last four and a half lines of Clause 2.3 are a hostage to fortune and would, in my opinion, be best omitted.

Assuming, however, all is sweetness and light, but that there are some instructions which in principle are non-contentious, i.e. both parties accept that they might or do actually move the twin goalposts of *cost* and *time*, what is the ACA Form provision for detailed agreement of such instructions? Clause 17.1 states:

"The Contractor shall . . . furnish the Client's Representative within 10 working days, or within such other period as may be agreed, . . . of receipt of the instructions with estimates of . . . "

(a) "The value of the adjustment (providing him with all necessary supporting calculations by reference to the prices contained in the Schedule of Activities) and the additional resources (if any) required to comply with the same and his method statement for compliance, and . . . "

(b) "the length of extension of time to which he may be entitled under Clause 11.5, and . . . "

(c) "the amount of any damage, loss and/or expense which may be suffered or incurred by him arising out of or in connection with such instruction, and . . . "

(d) "the necessary revisions (if any . . . to the dates and/or prices stated or contained in the Schedule of Activities."

Clause 17.2 assumes that such details can be agreed, but if not, Clause 17.3 provides that in the event of failure to agree:

" . . . all or any of the matters set out in them then:

> (a) the Client's Representative may never-the-less instruct the Contractor to comply with the instruction in which case the provisions of Clause 17.5 shall apply as if the Client's Representative had dispensed with the Contractor's obligation under Clause 17.1; or . . .
>
> (b) the Client's Representative may instruct the Contractor not to comply with the instruction, or . . .
>
> (c) the Client's Representative may refer the Contractor's estimates to the Adjudicator for his decision."

Clause 17.4 then provides that if the instruction is cancelled under Clause 17.3(b) the contractor has no claim, presumably to either the money or time, and Clause 17.5 allows the client's representative to make a unilateral "fair and reasonable . . . adjustment to the Contract Sum . . . and a fair and reasonable extension of time . . . ".

Finally, what if the contractor fails to comply with any of the pre-estimate provisions of Clause 17.1? Clause 17.6 provides the answer—or does it? Certainly the clause is a rare example, in the ACA Form, of convoluted wording, but broken down into manageable word bites, I understand Clause 17.6 to provide:

- assuming the client's representative has *not* dispensed with Clause 17.5 and pre-estimates under Clause 17.1 are still required
- then the client's representative is prohibited from certifying any value on interim valuation, but may do so on the final certificate only, and
- the contractor shall have no claim to interest thereon in the meantime.

As such Clause 17.6 is the equivalent of JCT 81 Supplementary Provisions S6.6 (Changes) and S7.6 (Loss and Expense), being a very real sanction on the contractor who plays fast and loose about providing pre-estimates under Clause 17.1.

For the purpose of all the ACA Form provisions a "working day" is defined at Recital G as a Monday to Friday, other than recognised public holidays in the country in which the works are to be executed and any day which is a holiday under the Building and Civil Engineering Annual Holiday With Pay Scheme from time to time in force, e.g. the traditional extra days at Christmas and New Year.

Summarising therefore the ACA Form provision for evaluation of employer's instructions under Clauses 8 and 17:

- There is a tight and businesslike time-scale framework, the parties being free to agree the actual periods prior to contract, but with a standard default provision of 10 working days for submission of pre-estimates by the contractor and five working days thereafter for agreement.
- In the event of non-agreement of cost or time the client's representative has three options which range from unilateral evaluation (subject to subsequent adjudication and arbitration), to immediate adjudication, to cancellation of the instruction.

Personally, I find the ACA Form Change Order procedure workmanlike and fair, allowing the client's representative to wield the big stick if necessary, but at the same time giving the contractor the necessary protection by way of adjudication if the client's representative abuses the power given him by the contract.

Delays

Clauses 11.1 to 11.4 set the basic ground rules:

- Once given possession the contractor shall complete the whole of the works in accordance with the agreed time schedule: Clause 11.1.
- If any section of the works are not deemed "fit and ready for Taking-Over by the Employer" at the due date "the Client's Representative shall issue a certificate to that effect": Clause 11.2.
- Given a certificate issued under Clause 11.2 the contractor shall allow the employer liquidated and ascertained damages at rates as prescribed in the time schedule, but the employer must authorise such deduction before "the Client's Representative may deduct such damages from the amount which would otherwise be payable to the Contractor on any certificate or the Employer may recover them from the Contractor as a debt": Clause 11.3.
- In the event that subsequently the employer fixes a later date for the taking-over of any section or of the works, or the adjudicator fixes a later date, the employer shall allow or repay the excess liquidated and ascertained damages, together with interest, to the contractor: Clause 11.4.

The foregoing are of course recognisable standard JCT standard provisions, but with two variations:

- The client's representative, subject to the employer's approval, does actually show deduction of liquidated and ascertained damages on the face of the payment certificate.
- The repayment of liquidated and ascertained damages is specifically stated as entitling the contractor to interest, i.e. the common law position, but now an express provision of the ACA/BPF contract.

My only criticism is that, having provided for the payment of interest, the contract could helpfully have prescribed a rate of interest, e.g. minimum lending rate plus 3 per cent.

As to the grounds for extension of time, Clause 11.5 starts from the position that "no extension of time shall be granted . . . ". There then follows the shortest list of exceptions, or grounds, of any standard contract: just five exceptions, of which the last, Sub-clause 11.5(e) will be the focal point for any design and build contractor:

"Any act, instruction, default or omission of the Employer, or of the Client's Representative on his behalf, whether authorised by or in breach of this Agreement."

As "instructions" are pretty well buttoned up both in respect of *cost* and *time* by Clause 17, the late-running contractor looking to create an extension of time and loss and expense claim is going to have to be very specific in establishing the Clause 11.5 ground on which his case is founded. Other than instructions under Clause 17 it is difficult to see much opportunity for the usual retrospective and speculative contractors' claims which waste so much of everyone's time and effort in our industry. A properly organised contractor knows within two weeks whether his intended progress is being, or is likely to be, disrupted by some "act, default or omission of the Employer". If he then fails to protect his contractual position by writing to the employer or his appointed representative flagging up the problem, the contractor only has himself to blame. If the problem is so flagged up, it can be openly debated—usually with the result that the problem can be worked out to everyone's satisfaction.

As such I believe the ACA Form with its tight requirements for the time schedule backed up by Clauses 11.5 and 17 is to be commended. These provisions are not in my view oppressive to the contractor as there is the safety valve of adjudication and arbitration, and above all else they provide for mutual certainty, i.e. both employer *and* contractor know at any point in the project where the twin goalposts of *cost* and *time* are set.

Finally I note the ACA Form expressly provides for acceleration and postponement at Clause 11.8, albeit the wording is clumsy. It would be a foolish client's representative who instructs the contractor to accelerate and seeks unilaterally to "ascertain and certify a fair and reasonable adjustment (if appropriate) to the Contract Sum . . . ". It would be far better, I suggest, if the client's representative adopted the alternative course provided in the second part of Clause 11.8 :

"Provided that if prior to giving any such instruction the Client's Representative requires the Contractor to give an estimate of the adjustment to the Contract Sum, the provision of Clause 17 (other than the provisions relating to extensions of time therein contained) shall apply as if an instruction given under this Clause 11.8 were included in Clause 17.1."

In other words, I would delete the middle sentence of Clause 11.8, leaving the change-order procedure as per Clause 17.

This welcome provision for the client's representative being empowered to instruct acceleration or postponement specifically relates to the changing of the previously agreed take-over dates. It does not restrict the client's representative's general power to admonish the contractor whom he perceives, or who is shown by analysis of the time schedule, to be running late. Therefore any such instruction to accelerate should be carefully worded and distinguished from a formal instruction issued under Clause 8.

Certification

Interim payments in the ACA Form are covered by Clause 16, Payment. The procedure is simply stated in that:

- "On the last working day of each calendar month up to and including the calendar month in which Taking-Over of the Works occurs . . .
- And thereafter as and when further amounts become due to the Contractor under this agreement . . .
- The Contractor shall present to the Contractor's representative an interim application . . .
- Stating the total amount due to the Contractor calculated in accordance with the provisions of Clause 16.2 . . .
- Supported by such documents, vouchers and receipts as shall be necessary for computing the same . . .
- Or as may be required by the Client's Representative."

These six essential elements of the contractor's obligation, and entitlement, are common ground to most standard forms of contract, noting the client's representative's power expressly stated in the last provision. In this last respect it is advisable for the client's representative to start as he intends to carry on, i.e. establish the level of detail and back-up required at the first valuation.

However, we then come to the more difficult part—bills of quantities or no bills? The ACA Form provides two alternatives, depending on whether the contract pricing document is:

- A priced Schedule of Activities—Alternative 1, or
- Contract bills—Alternative 2.

In either case, standard timescales are provided of 10 days from submission by the contractor for the client's representative to certify—but allows the parties to insert a different timescale if they so wish.

Under Alternative 1 the sophistication of the monthly valuation will depend on the detail contained in the elemental build-up shown in the priced Schedule of Activities, to which must be added a lump sum for "Preliminaries" and adjustment in respect of pre-agreed Instructions. Under Alternative 2 the procedure is the same, except that the assessment of work executed is more detailed, as allowed by reference to the contract bills.

In both cases the client's representative is expressly empowered to determine whether the work claimed for payment by the contractor has been "executed in accordance with this Agreement"—thus he may at his absolute discretion disallow work for payment purposes. There is then the usual provision for retention, the standard percentage suggested by the ACA being 5 per cent.

Thus the question arises—what is the status of the client's representative?:

- Independent certifier as the architect under JCT 80? Or . . .
- The employer's agent under JCT 81?

Given that there is the safety valve of adjudication the point is probably academic, but it is interesting to note that as compared with the express provisions of JCT 81, the ACA Form does not require the client's representative to give the contractor a statement setting out in detail the items or monies disallowed from the contractor's monthly application within a stated time-scale.

The final account provisions of the ACA Form are set out in Clause 19 and they follow the procedure of interim valuations, but with different time-scales, which again are subject to prior agreement between the parties:

- "The Contractor shall submit to the Client's Representative within 60 working days after the expiry of the Maintenance period the Contractor's final account for the Works . . .
- and all documents, vouchers and receipts as shall be necessary for computing the Final Contract Sum . . .
- or as may be required by the Client's Representative.
- The Client's Representative shall issue the Final Certificate within 60 working days after completion by the Contractor of all his obligations in accordance with this Agreement . . .
- The Final Certificate shall . . . from the 10th working day after the issue of the Final Certificate be a debt payable by the Employer or the Contractor as the case may be."

At best therefore the contractor is entitled to cry "foul" if he does not receive a final certificate within 120 days (60 plus 60, or whatever other periods are inserted in the contract) of the end of the maintenance period. However, this is subject to his completion of all other obligations, e.g. getting defects signed off and the delivery of all record drawings etc.

Hopefully, however, the whole final account process should be a straightforward bookkeeping exercise, reaffirming the penultimate certificate and releasing retention. Assuming Clause 17 has been fully operated there should be no room for contention. Accordingly, there is no reason why the parties should necessarily need the full 120 days to deal with the final account and conclude payment matters.

Certification of *quality* as built under the ACA Form is expressly reserved to the client's representative under Clause 16, as previously discussed under interim payments, and is implied under Clause 19.2 in respect of the final certificate.

Disputes

The ACA Form predated the Latham Report of 1994 and as such promoted the concept of adjudication as a first-step option for an aggrieved party to a construction contract, covering all potentially contentious issues, not just subcontract set-off disputes as previously provided for in the JCT Standard Green and Blue forms of subcontract.

At Clause 25.1 the ACA Form requires the adjudicator to be named, i.e. an individual chosen by the employer and notified to the contractor as part of the tender enquiry, or an individual mutually agreed by the parties prior to contract signature. Conceivably a firm or practice could be nominated at Clause 25.1 but I would want to see this qualified to only allow accredited adjudicators within that firm or practice to take the role and for such individuals to be identified prior to the commencement of the works.

In my opinion the ACA Form has got it right, i.e. it is infinitely preferable for the employer to select a respected and experienced individual, not otherwise involved in the project, than to leave the role open to institutional appointment, with the consequent twin risks of an inappropriate appointment and delay in getting the nominated adjudicator into post.

More importantly, the employer and his solicitors must guard against the very real potential for conflicting decisions down the contractual chain if different adjudicators are called in at different subcontract levels of essentially the same dispute—see Chapter 9 for detailed discussion of this important topic.

The referable matters set out at Clause 24.2 catch all potentially contentious post-contract issues, but only up to the taking-over of the works —leaving a very real "grey area", for example defects and their making good during the maintenance period, as discussed in Chapter 8.

However, what really excites me about Clause 25.2 is the provision immediately below the list of five referable matters:

- " . . . then such dispute *shall in the first place* be referred to and settled by the Adjudicator, . . .
- who, within a period of [5] working days after being requested by either party to do so, shall give written notice of his decision to the Employer, the Client's Representative and the Contractor . . . "

Please note the emphasis I have applied, in order to make it clear that the ACA Form is unique in that it makes adjudication a *compulsory* first-stage dispute procedure, without which there can be no right to arbitration under Clause 25.6, or litigation should Clause 25.6 be deleted.

The ACA's suggested time period of five days for the adjudicator's decision is laudable, but can of course be any period agreed by the parties. However, in practice it is wholly unrealistic for other than the most simple disputes, capable of instant inspection, i.e. "look-sniff" disputes—a quaint term from the days when arbitration was mainly concerned with the quality of perishable cargoes delivered to the London docks from around the British Empire!

A typical construction industry dispute will involve:

- Contacting the individual named in the contract as adjudicator, who may well be out of the country on business, if not on holiday;
- Confirming his immediate availability, which could be a good reason for not naming solicitors or known arbitrators who may well be committed to another case;
- Obtaining copy of the contract information relied on—Clause 25.2 not requiring such submission to accompany the Notice of Adjudication.
- Hearing what the *other* party has to say on the matter, including receiving copy of the paperwork they rely on—the most important duty of the adjudicator being to act, and being seen to act, even-handedly;
- Then considering the matter, with or without further questions, and deciding the issue, which involves drafting, typing and actually communicating the decision to all three parties.

As most disputes relate to payment of alleged monies due on interim valuation, clearly some period between 7 days and 30 calendar days is desirable, so probably the 28 calender days as required by the Housing Grants, Construction and Regeneration Act 1996 is a workable solution for most construction industry disputes. The exception is likely to be complicated design or extension of time issues, but at least the Act does allow the adjudicator the option of calling for another 14 calendar days, but only with the consent of the referring party. The above is a good example of how important it is to define "days"!

The ACA Form at Clause 25.3 expressly states that:

"the Adjudicator shall be deemed to be acting as expert and not as arbitrator . . . and his decision under Clause 25.2 shall be final and binding upon the parties until the Taking-Over of the Works . . . "

An arbitrator is immune from suit, i.e. action for negligence, unless he blatantly acts in bad faith, whereas an expert must exercise an appropriate duty of professional care in his advice or decision. Accordingly, an adjudicator appointed under the ACA Form may like to clarify his potential liability in his adjudicator's contract—assuming that he knows his name has been entered at Clause 25.1 and has had the foresight to agree his terms of appointment with the parties, albeit there is no dispute yet notified.

Quite what the position would be if the adjudicator is taken by surprise could be interesting. Assuming he drops everything and accepts the reference, how does he secure his terms and fee without delaying the process? Unfortunately the ACA Form makes no provision for the liability of the parties, either severally or jointly, for payment of the adjudicator's fee.

If the adjudicator refuses to act, or fails to act as required, Clause 25.4 provides a fall-back position, i.e. the parties apply to the President or Vice-President of the Chartered Institute of Arbitrators for the appointment of a replacement adjudicator.

Interestingly the ACA Form also expressly allows the adjudicator at Clause 25.1 to delegate "the performance of his duties" by notice to the three parties, i.e. employer, client's representative and contractor. This useful provision gets over the potential difficulties of:

- Non-availability, or
- The dispute being outside the skill-base of the named adjudicator.

The clause then provides an essential safeguard:

"The person so named shall have authority to perform the duties of the Adjudicator under the Agreement until such time as the Adjudicator shall notify the Employer, the Client's Representative and the Contractor from time to time that such person's authority is terminated."

Perhaps it might be simpler if the adjudicator named in the contract makes it clear in his notice of delegation that such delegation is for the specific single-issue dispute as currently referred, but understandably the standard clause has to be drafted on a default basis.

Clause 25.5 then warrants further emphasis:

" . . . if upon receipt of the Adjudicator's notice of his decision under Clause 25.2, either party is dissatisfied with the same *such party may . . . within [20] working days . . . give notice to the other requiring that the matter should be referred to the arbitration* of a person to be appointed under Clause 25.6 . . . *if no claim to arbitration has been notified by either party to the other within [20] working days as aforesaid, such decision shall remain final and binding upon the parties.*"

So now we come to Clause 25.6 and the ACA Form arbitration provisions, assuming that any contentious issue has survived the robust requirements of the preceding provisions of Clause 25. Whether or not the cut-off provision of Clause 25.5 fosters a culture of automatic Notices of Arbitration amongst "losing" parties to adjudication might be an interesting research question, but in any event such arbitration must be held over:

" . . . until after the Taking-over or alleged Taking-over of the Works or termination or alleged termination of the Contractor's employment . . . except with the written consent of the Employer and the Contractor."

So under the ACA Form the basic disputes regime is nothing but robust:

- First you *must* go to adjudication;
- Second, if you do not like the adjudicator's decision you *must* serve Notice of Arbitration within 20 working days;
- If you do not do so, your claim is lost.

What could be simpler?

As the icing on the cake, so to speak, Clause 25.6 tackles the key area where arbitration so often fails the parties, and which is the reason why arbitration clauses are often deleted from standard forms of contract, leaving disputes to be litigated in the High Court, namely joinder provisions.

Very rarely does one get a construction dispute which does not involve a third party, either up or down the contractual chain from the two immediate parties. The potential for "win–win" or "lose–lose" combinations of decisions reached in separate arbitrations essentially dealing with the same basic facts is not in the public interest, so the provision of Clause 25.6 in the ACA Form is especially welcome where one has a multi-party dispute. The provision is as follows:

"If in the Employer's opinion . . . any dispute or difference to be referred to arbitration under this Agreement raises matters which are connected with . . . matters raised in another dispute between the Contractor and any of his subcontractors or suppliers and . . . provided that such other dispute has not already been referred to an arbitrator, . . . the employer and the Contractor agree that such other dispute shall be referred to the arbitrator appointed under this Agreement, and . . . such arbitrator shall have power to deal with both such disputes as he thinks most just and convenient.

There is one problem with the above proposal—what if the subcontractor rejects the proposed arbitrator agreed between the employer and contractor?

The provision in Clause 25.6 could mean actually joining the two disputes, or structuring the hearings such that they interlock. This may not be easily achieved, but it can be effective given an arbitrator who is prepared to be the master of the proceedings. Alternatively, if "such other dispute" has already commenced, what is to stop the employer and contractor agreeing to appoint the existing subcontract dispute arbitrator? At least they will then be assured of a consistent result.

10.3   GC/WORKS/1 (EDITION 3)

General comments

The full title of this standard UK Government procurement agencies' contract is "GC/Works/1 (Edition 3) General Conditions of Contract for Building and Civil Engineering, Standard Form of Contract—Single Stage Design and Build Version".

This is but one of the government-promoted GC/Works/1 family of contracts, this version explicitly catering for the different allocation of risk and responsibilities inherent in design and build. This is achieved by following the structure and clause references of the lump sum version, but with amended clause wording where appropriate and with additional clauses where necessary, e.g. Clause 10a, Design Documents.

Very usefully it is prefaced by an "Introduction to the Conditions of Contract" which explains in a series of short paragraphs the fundamental requirements and key clauses of the design and build version.

As one might expect with a form of contract prepared by a client organisation charged with the protection of the public purse it is an unashamedly biased contract, giving the contractor virtually all the risks and responsibilities, and setting up a tight post-contract regime with virtually all power vested in "the Authority", i.e. the employer, and "the Authority's Project Manager".

Whilst being heavily loaded against the contractor, at least the ground rules are plainly stated in forthright terms and any contractor tendering for a Government or private sector contract on this particular form of design and build contract can price the risk loading accordingly.

Finally, at the back of the 65 standard clauses there is a helpful "Index Part 1—General" and an even better, "Index Part 2—Time Limits", summarising all obligations, whether on the contractor or the project manager, by clause reference.

There is then a two-page (actually $1\frac{1}{4}$-page) Model Form of Contract followed by a Model Form of Tender and a Model Form "Summary of Essential Insurance Requirements"—all very commendable for their brevity and relevance.

The "Introduction to the Conditions of Contract" stresses the need for formulating a contract strategy by a series of decisions which enable the allocation of " . . . design, construction, supply and commissioning responsibility between the various parties". This recognises that the authority's stated requirements might be expressed as a " . . . brief, describing basic requirements or a concept design for the Contractor to develop into a detailed design". I assume this means the brief could be written only, or it could be a schematic design, but in terms of allocating responsibilities, very few responsibilities remain with the authority once the contract has been placed.

The Introduction explains that the design and build version is specifically written for a single-stage design and build project and summarises the three criteria which determine when such a procurement is most appropriate:

- "When the Authority can precisely define its requirements."
- "When the Authority does not need ongoing control over the developing design."
- "When there is little likelihood of changes in requirements."

Taking these three criteria at face value one might struggle to think of the type of project which would so qualify, one possibility being a high-security prison not seen from any public area, but few other schemes would apparently pre-qualify on face value.

Rightly, the Introduction then warns that: "The Authority's key requirements must be clearly defined prior to signing the Contract, or else it must accept the Contractor's choices of materials and quality standards etc."

Benchmarking of quality

The Introduction advises users to fully specify materials and components which are " . . . important on either appearance or functional grounds" but otherwise recommends a performance specification. Surprisingly no mention is made of references to national standards, e.g. British Standards or Codes of Practice. Likewise no guidance is given as to benchmarking of required *quality* by the use of stating the required prime cost of materials, e.g. facing bricks or finishings and there is no guidance as to the use of provisional sums where the authority is unable to go firm on its requirements prior to tender.

Fitness for purpose

The Introduction, Item 14 gives the most lucid definitions of "Reasonable Skill and Care" and "Fitness for Purpose" one could wish to find, and then at Item 15 explains that the standard Conditions opt for the lesser of the two duties, i.e. "Reasonable Skill and Care" only. Item 15 goes on to advise users to use their judgement and that of their professional advisers as to whether the higher duty, i.e. "Fitness for Purpose" might be inserted, but warns that insurance cover might be a problem.

Frankly I find this advice ambivalent. If, as explained at the Introduction, Item 14, the statute law position under the Sale of Goods Act would be implied if the contract is otherwise silent, why not delete Condition 10(2)? The insurance market is very competitive, and always looking for new business. In recent years foreign insurers have entered the market, so why should the employer meekly give away the primary benefit of design and build, i.e. single-point responsibility, before the tender stage?

The bold approach is to make it a condition of a pre-qualification enquiry that "Fitness for Purpose" insurance cover for a stated amount can be obtained by each proposed tenderer. This soon sorts out the less reputable design and build contractors, particularly those who think design and build is just another version of that old game "pass the parcel", relying on the PI cover of outside design consultants.

Should this positive approach reduce the tender list, at least the employer has the early option of relaxing this requirement and obtaining tenders on the lesser basis of "Reasonable Skill and Care" only. In the event that the lower of the two lowest-priced tenders only offers the lesser design duty, there are then good grounds for considering the second tender on a perceived "Value for money" basis, rather than price alone.

Contract base-line

The top document is "the Authority's Requirements", as defined in Condition 1 and as expressly stated in Condition 2 as taking contractual precedence over "the Contractor's Proposals" in the event of subsequently discovered discrepancies. Condition 2 also provides for discrepancies within documents or with statutory requirements, planning consents, etc.

The third contract document is the "Pricing Document" which is undefined other than by way of the statement in Condition 3 that "this document replaces Bills of Quantities". The only further guidance on this essential contract document is given in the Introduction and advises that: "The Pricing Document will be used by the Authority in analysing the various offers made and to allow the Quantity Surveyor to evaluate change under Condition 42."

Introduction Note 16 further advises that: "It is important to have some means of pricing for any additional design costs that may be incurred by the Contractor when complying with changes resulting from a Variation Instruction." Otherwise Introduction Note 16 leaves it to individual quantity surveyors to draft the equivalent of the JCT 1981 contract sum analysis on a project-by-project basis and to incorporate this in the authority's requirements document. Frankly I find Condition 3 and Introduction Note 16 to be too generalistic and would have expected the standard contract as used by the UK's largest procurer of building works, let alone a committed "Best Practice" employer, to lead the industry by way of example.

Hopefully the next revision will see a Model Contract Sum Analysis incorporated by way of an optional Appendix at the back of the design and build version of the GC/Works/1 (Edition 3) Standard Form of Contract.

Finally in respect of the pricing document, condition 3(1) provides:

"There shall be no rectification of any errors or omissions in the Pricing Document nor any rectification of wrong quantities or wrong estimates of prices inserted therein by the Contractor or in any of his calculations or computations therein."

Presumably this Condition refers to post-contract procedure, i.e. the onus is on the authority's quantity surveyor to check and negotiate out any pricing anomalies *prior* to tender award. Indeed, reference to the Model Form of Tender enclosed at the back of the contract provides, at Item 5, that any obvious errors in pricing or arithmetic discovered on tender appraisal shall be referred back to the contractor.

Changes

Condition 1 and Introductory Note 17 then cover the role of the project manager (PM), the latter provision being commendably clear that, once the contract is awarded, the design and build contractor is in the driving seat as regards design development, viz:

"The primary role of the Project Manager on Design and Build contracts is inspection and quality overview, without muddling the design obligations of the Contractor. He is no longer responsible on behalf of the Authority for providing the Contractor with Design information required to enable him to construct the works."

Condition 10a then deals with the required procedure for the contractor obtaining clearance from the project manager as to his design development. All relevant design development work shall be the subject of a formally signed-off design document, the procedure for which is as follows:

10a(1) The contractor submits to the project manager two copies of the design document (usually drawings) before starting the work in question.

(2) The contractor shall not start work until he has ascertained that:

(a) "The PM has examined the Design Document and has confirmed in writing that he does not intend to raise any questions thereon, or . . .

(b) The PM has confirmed in writing that the questions he has raised about the Design Document have been answered to his satisfaction, and . . .

Upon completion of the procedures in (a) or (b) above the PM shall sign and date each copy of the Design Document, and return one copy to the Contractor."

No timescales are provided for the above procedure so it would be sensible to provide for such timescale in the authority's requirements document.

Condition 40, PM's Instruction, then distinguishes between "Instructions" and "Variation Instructions (VI)". In respect of "Instructions" as defined in Condition 40(2) the PM's power is fairly far-reaching, all "Instructions" needing to be in writing and with no provision for the contractor in any way challenging the PM's requirements.

Condition 40(4) implies that the PM must distinguish "Variation Instructions" from "Instructions", the procedure in respect of the former being that the contractor may be instructed to submit a written lump sum cost quotation to the quantity surveyor within 21 days, including advice as to the likely effect on programme. Conditions 41 and 42 then provide change valuation and procedure rules, including what happens in the event of failure to agree the costs involved. In this event Condition 42(4) provides that "the PM shall instruct the QS to value the Variation Instruction", subject to the usual pecking order of contract pricing, fair rates, dayworks.

The QS must then prepare his valuation and notify the contractor within 28 days, following which the contractor can submit his own reasoned valuation within a further 14 days—Conditions 42(8) and (9); "Days" are usefully defined in Condition 1 as "calendar days".

What then happens if the two parties still cannot reconcile their cost assessment of a "Variation Instruction" is left hanging—very unhelpfully. The only possible escape route is that provided by Condition 42(5)(c), i.e. the QS agrees that the *direct* cost of the work is recorded on a daywork basis, but this begs the question of indirect costs, such as disruption and additional design. In the meantime the unhappy contractor is stuck with Condition 40(3), i.e. "The Contractor shall comply forthwith with any Instruction" which, according to Condition 40(1), includes "Variation Instructions".

Delays

Condition 33(1) states the usually recognised objectives of a contractor's programme, but in addition requires the contractor to state:

- The day or days on which possession of the site or parts of the site are required;
- The details of any temporary work;
- His proposed method of work;
- Labour and plant proposed to be employed;
- Events which in the contractor's opinion are critical to the satisfactory completion of the works.

Condition 33(1) is, however, silent as to whether the submission of such a programme is to be a pre- or post-tender requirement, but the Introductory Note to Condition 33 recommends that the authority's requirements document calls for programme submission with tender.

Assuming this advice is followed the contractor's programme will be part of the contractor's proposals and so qualify as a contract document. That this is the intended approach is further confirmed by Condition 1 where "the Programme" is defined as:

"The document or documents submitted prior to acceptance of the tender and agreed at that time by the Authority or amendments thereto agreed by Contractor and the PM in accordance with Condition 33 (Programme)."

Inevitably some refinement of the contractor's programme may be desirable or necessary between tender and contract award. In this event any amendments can be mutually agreed and the programme revised before contract signature.

Once so incorporated in the contract documentation the contractor's programme will be a key monitoring tool as well as a base-line for assessing the *time* effects of any variation instructions.

Condition 35 then formalises, as a contract requirement, the generally accepted industry procedure in respect of progress meetings:

- Regular site meetings, usually monthly, to be called by the project manager;
- Contractor to submit a written report five days before the site meeting covering design progress and construction progress against programme, as well as listing new circumstances which have caused or might cause delay, any new applications for extensions of time and any reprogramming proposals.

The Project Manager then has seven days after each site meeting to issue to the contractor a written statement confirming his opinion on:

- The current works progress position as compared with the current contractor's programme intention;
- What matters have or are likely to delay completion of the works or any section thereof;
- Any acceptance of the contractor's reprogramming proposals;
- A summary of position of applications and awards for extension of time under Condition 36.

The above provisions of Condition 35 make eminently good sense and could usefully be considered by the various committees charged with production of all future revisions to the various standard forms of contract from the JCT downwards.

Condition 36 then buttons up the formal procedure for the contractor making an extension of time application, noting that the project manager may initiate the procedure. Condition 36(2) sets down the grounds on which extension of time may be founded, namely:

(a) a change in the authority's requirements, i.e. a variation instruction;
(b) the act, neglect or default of the authority or the PM;
(c) Any strike or industrial action outside the control of the contractor or any of his subcontractors;
(d) An accepted risk or unforeseeable ground conditions—the former being listed in Condition 1, Definitions etc.
(e) Any other circumstances (other than weather conditions) which are outside the control of the contractor or any of his subcontractors and which could not have been reasonably contemplated.

This last ground is particularly interesting in that it gets away from the arbitrary requirement of the JCT provisions for the architect or employer's agent to decide what is, or what is not, "exceptionally inclement weather" this event being frequently abused to let in extensions of time for which the design team are arguably responsible, but which at the same time deny the contractor reimbursement for the costs of extended preliminaries.

Quite what would happen to contractors under the GC/Works/1 if we had a repeat of the winter of 1963, when most sites shut for two months, would be presumably at the discretion of the employers. On a site outside Manchester I witnessed the contractor trying to excavate a spoil heap in the first week in May and the bucket banging on solid ice once the first 0.5 m had been removed!

Condition 36(3) specifically allows the project manager to give an interim decision on the contractor's extension of time application and again I consider the JCT might do well to adopt this open approach.

Condition 36(4) provides the usual cut-off for extension of time claims at completion of the works and requires the project manager to finalise all discussions in respect of extensions of time within 42 days of completion of the works.

Following the FIDIC example, Condition 36(5) provides for a dissatisfied contractor making a detailed application to the project manager within 14 days of a decision given under Condition 36(1) asking for a re-decision. The project manager then has 28 days to notify the contractor of his further decision. Quite what the position is should the project manager fail to give such decisions within the required timescales is unstated, but I note that Condition

36(6) puts the common law duty on the contractor to mitigate loss, i.e. to minimise any delays where possible.

In this context it is to be assumed that given an awarded extension of time under Condition 36 the project manager can call for acceleration proposals from the contractor under Condition 38. In all three sub-clauses the drafting is clumsy, in that in the section "If the Authority (1) wishes . . . (2) accepts . . . or (3) considers . . . *he* shall . . . " the project manager is not mentioned, but presumably is the person or body referred to as "he".

Such explicit provision for acceleration is again commendable, but in the event that terms for such acceleration cannot be agreed it would seem that the contract falls short of giving the project manager power to formally instruct. As such this would appear to be one of the few situations under the contract where the contractor has the better bargaining position.

Certification

As was made clear in respect of the project manager's role in Introductory Note 17 his primary role is inspection and *quality* overview. Also, as the personal representative of the authority, the project manager is the individual responsible for effecting all obligations ascribed by the contract to the authority. Specifically the project manager is responsible for certifying extensions of time under Condition 36 and for certifying completion under Condition 39.

However, in respect of money matters the primary role is given to the quantity surveyors, e.g. valuation of variation instructions under Condition 42 and the final account under Condition 49. Strangely the quantity surveyor gets no mention in respect of interim payments, which at Condition 48 are rather quaintly titled "Advances on Account" and which are payable by the authority under Condition 50. The interim payment or "Advances on Account" procedure of GC/Works/1 (Edition 3) is prefaced by the condition that "any relevant Instructions have been or are being complied with", i.e. the authority can, *in extremis*, deny the contractor a monthly valuation on a relatively minor technicality.

Hopefully this very real power is not abused, but I can certainly think of one or two instances in my 30 years' experience when such a powerful sanction would have been the only way to make a defaulting contractor comply with essential provisions of the contract, e.g. production of concrete cube test results or fire door certificates!

Otherwise the procedure shall be as follows:

"(a) 95% of the proportion of the sum specified as the proportion of the Contract Sum for the relevant month according to the Stage Payment Chart or Charts.
(b) 100% of any amount agreed under condition 42(I)(a) (Variation of Variation Instructions) in respect of work completed in the relevant month.
(c) 100% of the agreed value, or, failing agreement 95% of the QS's valuation, under Condition 42(5) (Valuation of Variation Instructions) and Condition 43 (Valuation of other Instructions) in respect of work completed in the relevant month.

(d) 100% of any amount determined by the QS under Condition 46 (Prolongation and Disruption) in respect of the relevant month.

(e) 100% of any amount calculated under condition 47 (Finance Charges)."

The above wording is hardly an exercise in plain English but when read and cross-checked with the clause references it does make sense. However, being based on pre-agreed stage payments it does assume that the contractor is running on time.

The very real possibility that the contractor might not be performing according to his own programme is then protected by Condition 48(3) which provides:

"Where the P.M. has recorded in a statement after a progress meeting that the Works are in delay or are ahead of programme he shall by reference to the Stage Payment Chart or Charts adjust the Contractor's entitlement to payments in accordance with paragraph (2)(a)."

The remaining point of interest in the GC/Works/1 (Edition 3) interim payment provisions is Condition 48(5) which gives the PM the right to require the contractor to evidence payment of previously certified amounts in respect of subcontractors and suppliers.

The final account procedure under Condition 49 then puts the onus on the authority's quantity surveyor, subject to the satisfaction of the project manager, to prepare the final account and forward same to the contractor. Quite why it is envisaged it might take six months to tot up the net cost of pre-agreed variation instructions, together with the re-assessed value of those variations where costs are disputed, and add same to the contract sum, I do not know.

Condition 49 provides for the contractor then having a further three months to signify his acceptance of the authority's draft final account, or alternatively to reject same and provide his own detailed final account. Here the position is lax as compared with JCT 80 where the contractor must register his objection if the architect's final certificate is not to become binding and conclusive. What is the position under GC/Works/1 (Edition 3) if the contractor neither agrees or disagrees within the prescribed three months, I ask? The contract is silent.

Finally in connection with certification of monies it is worth noting that the design and build version of GC/Works/1 provides at Condition 47 for finance charges on any late payment by the authority to the contractor, subject to certain safeguards—yet another commendable feature which the JCT drafting committee would do well to consider.

Disputes

As for the "hard" version of JCT 81 With Contractor's Design, the design and build version of GC/Works/1 has a two-stage disputes procure. The first stage, provided by Condition 59, is adjudication, subject to three provisos:

(a) Adjudication is only available up to certified completion of the works.

(b) The dispute must have been ongoing for at least three months.

(c) Certain matters are specifically stated in the contract as to be final and binding:

- Condition 18, Measurement;
- Condition 26, Site Admittance;
- Condition 31, Works Quality;
- Condition 39, Certifying Work;
- Condition 40, PM's Instructions;
- Condition 44, Labour Tax;
- Condition 56, Determination

Disputes in these areas are *excluded* from adjudication—subject to the provisions of the Housing Grants, Construction and Regeneration Act 1996.

The GC/Works/1 (Edition 3) appointing procedure for the adjudicator under Condition 59 is somewhat strange, but is as follows:

- The contractor serves a fully detailed written notice on the person named in the Abstract of Particulars as the person to whom requests for adjudication should be referred.
- This person then nominates an officer of the authority, or a person acting for the authority, who has not previously been involved with the project to act as independent adjudicator—and presumably passes the contractor's written submission to him.
- This same person as named in the Abstract of Particulars also informs both the contractor, the project manager and the quantity surveyor of his nominee to act as the adjudicator.
- The project manager and the quantity surveyor then have 14 days to submit representations to the nominated adjudicator from the date of receipt of the Contractor's Notice.
- The adjudicator is then required to notify his decision, with or without reasons, to the parties within 28 days of receipt of the Contractor's Notice, and to allocate his fees.
- In reaching his decision the adjudicator shall have regard to how the parties have complied with the contract rules and whether they have acted in good faith.
- The decision of the adjudicator is binding until completion—whereafter arbitration is available to either party, if not deleted from the signed contract.

Whether or not the nominated adjudicator is so nominated on a reference-by-reference basis, or once nominated remains the adjudicator for all future disputes is not covered by Condition 59. For the avoidance of doubt this needs to be clarified in the contract documentation.

Condition 60 then provides that, unless otherwise agreed, no arbitration may be commenced until completion or abandonment of the works, or the determination of the contract. Further, the right to arbitration is expressly excluded in respect of those same matters as those excluded from adjudication.

As to the appointment of the arbitrator, this is triggered by notice of arbitration served by either party on the other inviting the other to agree the name of a single arbitrator. Failing agreement as to the name of the arbitrator "within a reasonable period" (undefined), only the authority can apply to the Chairman or Vice-President of the Chartered Institute of Arbitrators for an independently appointed arbitrator. In practical terms this condition is a nonsense—what if the authority simply refuses to agree an arbitrator? The contractor could then be stuck, as he has no right himself to apply to the Chartered Institute of Arbitrators.

Assuming the authority does not obstruct the process, Condition 60 then continues with the arbitration procedure to be followed once an arbitrator is appointed—this requires no further comment.

10.4   THE IChemE MODEL FORM OF CONDITIONS OF CONTRACT FOR PROCESS PLANTS

General comments

The IChemE have issued two standard Forms of Contract:

- The Green Book—applicable to cost reimbursable contracts.
- The Red Book—applicable to lump sum contracts.

Both of these standard forms are necessarily design and build contracts, as standard practice in the process engineering industry is for the employer, i.e. the plant operator, to specify his requirements—in detail or in outline—and for the contractor to take over from that point once the details of design, and end-product output, have been agreed and converted into a contract where *cost*, *time* and *quality* are defined.

Accordingly the Red Book contract is the nearest equivalent of the IChemE's two Model Forms to the building industry's JCT 81 With Contractor's Design. As such, any comparison of the IChemE Form with JCT 81 is interesting in that the employer's objectives may be differently prioritised, but the core issues should represent common ground. The starting point is therefore the Introductory Notes issued with the IChemE Red Book Form of Contract.

The Introductory Notes in the Red Book stress the need for time to be taken in " . . . gathering information and making many decisions based on the preparation and discussion of drafts" and for the scope of the required work to be " . . . clearly defined beforehand".

A process plant project may well involve all or some of the following activities:

- Comparative evaluation of different processes;
- Chemical engineering design;
- Detailed engineering;
- Supply of hardware;
- Site construction work;
- Starting up, testing and plant operation.

A process plant contract may be formed by the joint signing of a formal negotiated contract or by exchange of tender enquiry and written acceptance; the IChemE Model Form is in two parts:

- General conditions of contract, supported by
- Schedules and Special Conditions of contract.

Since this is an evolving design process focused on the end product—and the ongoing performance of the manufacturing plant—design responsibility is handed over by the employer to the contractor upon acceptance of tender. Specifically, the Red Book Standard Form of Agreement, Appendix 1, Second Recital requires the contractor to "complete such design and to execute and . . . complete the Works". Should a tenderer have any difficulty in fully meeting the employer's requirements, any problems or alternative solutions must be identified by the tenderer in his submission.

Benchmarking of quality

Clause 20 of the IChemE Model Form, General Conditions, provides a detailed procedure for the contractor's post-contract design development, stating timescales and rights of rejection by the engineer on *two* grounds only:

- Non-compliance with an express term of the contract;
- The contractor's proposal is contrary to good engineering practice.

The Special Conditions will cover the particular details of the project in question, e.g. the nature of the process, any legalities such as ownership of the process and patents which must be safeguarded, and security arrangements —all of which the contractor will be required to take into account in his design development.

Any particular standards and *quality* required should also be covered by the Special Conditions or the Schedules by expressly stating the detailed requirements, or by specific reference to nationally accepted standards or the Codes of Practice.

Fitness for purpose

Fitness for purpose is implicit in any process plant contract, as is made clear in the Introductory Notes which list some key criteria which may well be involved:

- Quantity and *quality* of the required manufactured products;
- By-products and effluents made by the plant when it is supplied with specified raw materials, and utilities;
- Product yields and efficiencies or their equivalent;
- The utility consumption, i.e. raw material input;
- Any other appropriate chemical engineering criteria.

Contract base-line

The IChemE Model Form of Agreement Red Book envisages the contract documentation consisting of:

- The Form of Agreement
- The General Conditions of Contract
- The Special Conditions (if any)
- The Specification and Drawings (if any) or annex thereto
- The following Schedules:

Schedule 1: Description of Works
Schedule 2: Drawings for Approval
Schedule 3: Final Drawings and Manuals
Schedule 4: Times and Stages of Completion
Schedule 5: Take-over Procedures
Schedule 6: Performance Tests
Schedule 7: Payment of the Contract Price
Schedule 8: Liquidated Damages:
    Part I for Delay
    Part II for Failure of Performance Tests

As regards definition of the contract price, Schedule 7 is the key document and it is left to the parties as to the level of detail disclosed therein relevant to the Description of Works in Schedule 1.

Changes

Given that the IChemE Red Book is a lump sum form of contract, provisions are made for the administration of variations or changes as follows:

Clause 16:   Allows the employer through his appointed engineer to instruct a variation, with or without having first obtained a quotation from the contractor, and at the same time it seeks to protect the contractor from major disruption to the normal course of his business by providing a cost cap of 25 per cent (unless otherwise agreed) to the cumulative value of additional work.

Clause 17:   Allows the contractor to put forward a variation for consideration and rejection/adoption by the engineer.

Clause 18:   provides the basis of cost adjustment to the contract price for variations—either by prior agreement, by reference to a Schedule of Rates, or by third party determination (see later comments).

The need for fair evaluation of variations, or changes, is recognised by the "Guide Notes for the Preparation of the Schedules and Special Conditions of Contract" as bound into the back of the IChemE Model Form of Conditions, item AA. Essentially, it is recommended that the parties agree a Special Condition specifying the manner of pricing variations, and an extra Schedule setting out agreed rates for labour, materials and equipment .

Interestingly, the Guide Notes then recommend that:

"Virtually all the factors that would affect the price of a reimbursable contract need to be considered in reaching a fair method of pricing a Variation in a lump sum contract, but in many cases these can be accommodated in a simple Schedule by incorporating profit, overheads, payroll burdens and the like into the rates."

All this is further explained in the practice notes accompanying the IChemE Model Forms of Conditions Green Book for Cost Reimbursable contracts.

The question of adjustment of design fees would appear to be left at large, unless subsumed by "payroll burdens".

The IChemE Model Form of Conditions, Introductory Notes, suggest that a project co-ordination procedure covering such matters as day-to-day communication between the parties, working methods and programme could usefully be included in any process plant contract, which, in respect of design development and drawing approval by the engineer, shall be read in conjunction with Clause 20.

Delays

Clause 14 of the IChemE Model Form of Conditions Red Book provides the procedures for the administration of delay claims by the contractor. It is very similar to the provisions of JCT 80 and 81 in that the contractor must give notice to the engineer, subject to six possible grounds listed at Clause 14.2. However, the engineer must then decide and give notice of his decision to both the employer and the contractor. Unlike the standard FIDIC provision there is no provision for the matter being referred back to the engineer for a second opinion should the contractor not be too amused by the engineer's first decision.

The concept of liquidated damages is provided for at Clause 15, but it is an option usually capped to an agreed percentage of the contract price, to be entered into the contract. Otherwise unliquidated damages will apply, which in process plant terms could be many times greater than the contract price.

Certification

Unlike JCT 81 With Contractor's Design, the IChemE Model Form of Conditions Red Book does have an independent certifier in that this duty is given by the contract to the engineer, albeit he is appointed and paid by the employer, otherwise known in Red Book terms as "the Purchaser".

The actual payment procedure is covered by Clause 39, which allows the contractor to submit regular invoices to the engineer for payment according to the period set out in Schedule 7. The engineer then has seven days to issue a certificate to both the employer and the contractor detailing how the payment is made up, and including either agreed or assessed amounts for variations.

The employer then has 28 days to make payment, following which the contractor:

- Is entitled to interest on a daily compound basis at 12 tenths of the Bank of England minimum lending rate, and . . .
- Is entitled to give notice to the employer that if after a further 14 days the required payment, plus interest, is not received then he reserves the right to suspend all work until such time as full payment is received, and . . .
- Is entitled to adjustment of the contract price in respect of the additional costs of any supervision and to any consequential effects, e.g. a revised completion date.
- If after 120 days the works remain suspended the contractor may serve notice on the employer and the contract shall be treated as if determined by the employer (purchaser).

In a process plant contract, performance testing is the real test of practical completion. Accordingly the JCT terminology of "practical completion" is replaced in the IChemE Model Form of Conditions Red Book by reference to "taking over" (Clause 34) and "performance tests" (Clause 35).

Seemingly there are no contractual provisions for presentation and agreement of the final account, but there is a draconian provision for issue by the engineer of the final certificate (Clause 38). When *all* tests and *re-tests* have been completed and any defective work has been made good and tested again, the stated defects liability period runs afresh and there can be no final certificate, until the works production has been trouble free for the full defects liability period since the last remedial works—fine words, but what constitutes remedial works as opposed to maintenance and fine-tuning could be hotly contested.

By implication therefore, the contractor cannot expect to agree his final account with the engineer *and* be paid by the employer until the latter is entirely satisfied with the performance of the plant!

Disputes

The IChemE Model Form of Conditions Red Book has a two-part disputes procedure for certain issues only, e.g. as we have seen in the matter of delays and extension of time, the engineer's decision is not subject to review or second opinion.

Clause 47 provides for arbitration generally in respect of all matters arising under the contract *except* those matters which are referable to an expert under Clause 46.

Such "expert" matters listed under Clause 46 which are specifically *not* referable to Arbitration are:

Clause 16.6:   Valuation of variations;
Clause 16.8:   Contractor's refusal to undertake a variation;
Clause 17.2:   Whether a contractor's proposed variation concerns the elimination of a specified hazard;
Clause 20.5:   Engineer's disapproval of the contractor's drawings.

Again there is an exception to the general rule, i.e. that matters specifically reserved for resolution by the expert cannot then be referred to arbitration, and this is:

Clause 42.11:   Termination of the contract (seemingly with arbitration option as it is not reserved for the "expert" under Clause 46).

The appointment procedure under the IChemE Model Form of Conditions Red Book for both experts (Clause 46) or arbitrators (Clause 47) is by mutual agreement, failing which one party applies to the President of the Institute of Chemical Engineers, who duly appoints.

However, given the introduction of the Housing Grants, Construction and Regeneration Act 1996, introducing compulsory adjudication on all projects which fall within the definition of construction works, it must be assumed that the IChemE Red Book will be amended to bring in the "adjudicator".

10.5 THE ENGINEERING AND CONSTRUCTION CONTRACT, NOVEMBER 1995

General comments

Originally entitled the "New Engineering Contract" (or NEC), this Institute of Civil Engineers sponsored initiative represents a worthy attempt to start with a clean sheet of paper and to create a new standard form of contract, applicable across the building and civil engineering industries, where the contracting parties are able to resolve their disputes from a level negotiating position with or without the assistance of a third-party adjudicator.

Now known as the Engineering and Construction Contract (or ECC), the original 1991 NEC version was identified in the Latham Report of 1994 as the preferred way forward for the industry. However, the take-up of the ECC in preference to other proven forms of standard contract has been very limited, probably simply owing to the old preference for the devil you know rather than the devil you don't know, but also because it clearly needs further work and "road-testing".

Not surprisingly, the ECC follows the basic two-part structure of the well-known FIDIC contract arrangement of General and Particular clauses—now known as "Core" and "Optional" clauses.

In this neat and structured way all usual procurement options are catered for, i.e. the whole spectrum of risk allocation between employer and contractor as represented by lump sum contracts, target cost contracts, cost reimbursable and management contracts.

Making up the core clauses are contract provisions common to all forms of procurement, such as:

- Names of the parties;
- Definition of contract terms;
- Duties of the named players, i.e. employer, project manager, supervisor, contractor, subcontractor and adjudicator;
- The contractor's main responsibilities;
- Time provisions;
- Testing and defects;
- Payment provisions;
- Variations—known as "compensation events";
- Title, i.e. ownership of materials and plant;
- Insurances;
- Disputes and termination.

These "core" clauses are followed by six main options:

A. Priced contract with activity schedule;
B. Priced contract with bill of quantities;
C. Target contract with activity schedules;
D. Target contract with bill of quantities;
E. Cost reimbursable contract;
F. Management contract.

There then follow the secondary options G to Z which sweep up all other matters as a potential shopping list, some of which will be relevant and others not.

The essential structuring of the ECC core clauses divides them into nine sections:

1. General;
2. The contractor's main responsibilities;
3. Time;
4. Testing and defects;
5. Payment;

6. Compensation events;
7. Title;
8. Risks and insurance;
9. Disputes and termination—with the provision for adjudication as a first-stage optional process.

Quite why the actual clauses under each of these nine sections are then numbered 10, 20, 30, etc. is a mystery which the explanatory notes given under "Clause Numbering" in the "Guidance Notes" do nothing to dispel.

The design and build version of the ECC family of contracts is then formed by a framework combination of:

* The *core clauses*
  +
* Option A: Priced contract with activity schedule
  or
* Option B: Priced contract with bill of quantities
  +/−
* Option M: Limitation of contractor's liability for his design to reasonable skill and care
  +
* *Data provided by the employer* wherein the design responsibility is allocated fully to the contractor, and the employer's requirements are set out by way of a performance specification supported by schematic design drawings, if required, as part of the "Works Information".

Of the *core clauses*, Clause 21.1 is the key clause, requiring the following to be stated in the employer's "Works Information" document:

* The amount of design to be carried out by the contractor—whether partial or total;
* The required criteria for the contractor's design—whether form, geometry, dimensions of the works, specifications, codes of practice, standards and environmental criteria;
* Any limitations which the employer wishes to impose upon appearance, durability, operating and maintenance cost, etc.

Otherwise the EEC Guidance Notes and Flow Charts are particularly unhelpful in respect of how to procure a design and build project, and probably deliberately so, as the whole principle of the EEC is that it is a common form of contract applicable to procurement of all building and civil engineering projects.

As such the traditional divide between JCT 80 Employer's Design, JCT 81 With Contractor's Design and JCT 87 Management Contract disappears—the ECC is one standard contract—but with significant options in the filling to the pie, so to speak, to bring out the required allocation of the traditional risks of design and construction, not forgetting supervision.

Benchmarking of quality

Beyond the recommendations of the ECC Guidance Notes concerning Clause 21.1, already referred to, the ECC does not seek to tell the employer how he should state his requirements and express his expectation of finished *quality*. This is left to the employer and to the text of the "Data provided by the employer", but it is interesting to note the comments of the Guidance Notes in respect of Clause 21.2:

> "Sometimes the Project Manager will see in the design submitted by the Contractor characteristics which, if they had been foreseen, he would earlier have stated to have been unacceptable by including an appropriate constraint in the Works Information."

The ECC Guidance Notes go on to explain that in this event the project manager may then impose such a constraint, i.e. post-contract, but that this will be a "compensation event".

So far so good, but then the real problem is how to value the constraint. A commercially aggressive contractor will doubtless play the game of offering minimal *quality*—and when the project manager rejects it, but cannot positively show that this is below the standard identified in the "Data to be provided by the Employer" and the "Works Information", the contractor will claim a "compensation event". This obvious area for disgreement therefore needs particular attention by the employer and his team when setting up the contract, i.e. benchmarking of *quality* is a key pre-contract area.

Fitness for purpose

The preferred philosophy of the ECC appears to be that the employer should state in the contract data the financial limit up to which he requires the successful tenderer to insure against failure of that tenderer's subsequent design development and final delivery. The phrase "fitness for purpose" does not appear, and so I suspect the ECC is thinking in engineering terms, i.e. physical failure of structure, rather than the JCT 81 concept of "fitness for purpose" in its wider application.

The relevant *core clause* is Clause 21.5, which is deceptively simple, giving two options should notification not be viable for any reason:

- State the limit of liability for design defects not listed on the Defects Certificate, i.e. latent defects, typically £250,000 limit; or . . .
- Exercise Option M limiting the design and build contractor's obligation to reasonable skill and care only.

Personally I believe that the employer's best interests are served by *not* giving away design and build's unique selling point, i.e. single-point responsibility, and that contractors with a good claims record will not find it too difficult to obtain fitness for purpose insurance cover in their own name, subject of course to subrogation of design liability if they are delegating to an outside consultancy.

I would therefore favour inserting a specific provision for fitness for purpose in the "Data to be provided by the employer" text and inserting "unlimited" in the Appendix against Clause 2—a typical Appendix for a mythical project having been set out by way of example at Appendix 5 of the ECC Guidance Notes. Which leads me to a very real complaint: I know that they are printing train timetables smaller and smaller, such that they are illegible to all but those blessed with perfect vision, but I see no reason why the ECC should follow suit! Please, at the next reprint, may we have a larger typeface for these Appendices?

Contract base-line

As previously stated the build-out process will be commenced at tender stage by the issue of a "Works Information" document setting out the employer's requirements which will include directly, or by way of a separate document, the details to be inserted in the *core clauses*. Thus a typical enquiry package will include schematic drawings and a performance specification identifying the scope and *quality* of the work required.

If the employer wishes to go for single-point responsibility in respect of design, Option M will be left on file and not used—and vice versa, if the employer goes for the lesser duty. As the Guidance Notes explain:

"Without this option the Contractor's liability for his design is strict, that is, it must be in accordance with the Works Information. This option reduces his liability for his design to 'reasonable skill and care'. In any dispute, the burden of proof is on the Contractor to demonstrate that he used reasonable skill and care, not upon the Employer to show that the Contractor did not."

In terms of base-line documentation, this then leaves the pricing document, which as we have seen is always the most difficult area to pin down when setting up a design and build project.

So we come to the A or B options of the basic ECC contract framework —priced contract with activity schedule, or priced contract with bill of quantities, the former being a rather blunt tool, but one more familiar to civil engineering quantity surveyors and the latter being "bread and butter" to the building industry quantity surveyor. As previously discussed, whichever option is specified in the "Instructions to Tenderers" the employer must lay down the ground rules and detail required if the contractor's submission is to be useful for *both* interim valuations *and* assessing the proper cost of "compensation events".

If Option A—priced activity schedule is chosen, the Guidance Notes make it clear that the activity schedule must not be used to describe the works, but the employer may state certain "activities" which he requires the contractor to state and presumably price. Otherwise it is left to each design and build tenderer to formulate his own scheme design and associated "Schedule of Activities", broken down into groups of activities which should be capable of easy site identification.

Given that the entitlement to be paid only arises when each group of activities, priced on a lump sum basis, has been completed on site, it pays a contractor to break down his chosen activities into reasonable detail, rather than group them together in larger lump sums.

Finally, under Option A, the Guidance Notes make the point that design and temporary works activities should be separately identified by the contractor in his "Schedule of Activities" if he requires these to trigger payment entitlement.

The alternative, Option B—priced bills, is familiar territory, but the Guidance Notes are less than clear when implying that the bills of quantities will be an employer-prepared document, issued as part of the tender enquiry, stating the method of measurement used, and any deviations therefrom. This concept can only assume the scheme is more or less fully designed by the employer, unless it is a tender requirement that each tenderer produce a short form bill of quantities quantifying his proposed design.

Thus a typical ECC tender enquiry package on a design and build project might consist of:

- Instructions to tenderers, including instructions for preparing the Activity Schedule under Option A;
- A form of tender;
- Contract data—Part 1;
- Contract data—Part 2 (proforma for completion and return by tenderers);
- Bills of quantities, under Option B;
- Works information;
- Site information.

Changes

Changes, alias "compensation events" under the ECC, are defined by Core Clause 60.1 and run to 18 situations. All of these are recognisable from other contracts, the only real point of interest being the ECC attempt to deal with the problem area of weather. The ECC approach is to require that:

"A weather measurement is recorded
- Within a calendar month
- before the Completion Date for the whole of the works, and
- at the place stated in the Contract Data
the value of which by comparison with the weather data, is shown to occur on average less frequently than once in ten years."

In other words, even the hurricane of 1987 would have been debatable—as we had another one only slightly less violent a year later!

Clause 61 deals with notification requirements for claimed "compensation events", distinguishing between:

- Events which have arisen by virtue of instructions given—(Cl. 61.1);

- Proposed instructions or proposed changes of decision—(Cl. 61.2);
- Events which have happened or which may yet happen as a consequence of other factors which may qualify as a compensation event—(Cl. 61.3).

The presumption is that the project manager will call for quotations, which under Clause 62.1 cover both *cost* and *time*.

Clause 61.4 gives the project manager total authority to decide the merits or otherwise of the contractor's assessment of a compensation event, and Clause 62.3 provides that the contractor has three weeks to provide any requested estimate, whereafter the project manager has two weeks to decide thereon, which can take the form of:

- An instruction to submit a revised quotation;
- An instruction accepting the quotation;
- A notification withdrawing the proposed instruction or changed decision;
- A notification that the project manager will be making his own assessment.

There then follow at Clause 63.1 the rules for evaluation of compensation events which include the concept of:

- "Actual Cost of work already done";
- "The forecast Actual Cost of work not yet done"; and
- The resulting fee.

If one then turns to the Definitions at Clause 11 one finds "Actual Cost" defined as " . . . the cost of the components in the Schedule of Cost Components whether work is subcontracted or not excluding the cost of preparing quotations for compensation events".

On page 37 of the ECC one finds a three-page "Schedule of Cost Components" which equates to dayworks, followed by a two-page "Shorter Schedule of Cost Components", which is an option under Clause 63.11. The procedure is completed by Clause 64 summarising in what situations the project manager assesses a compensation event in terms of *cost*, (Cl. 64.1) and in terms of *time* (Cl. 64.2), notifying his assessment (Cl. 64.3), whereas Clause 65 covers the implementation of compensation events.

All the above seems to be a tortuous reinvention of the wheel—and fertile ground for disputes between the project manager and the contractor.

No express provision or cross-reference is made in Section 6, alias Clauses 60 to 65, to adjudication, but this is covered by *Core Clause* 9 and specifically Clause 90.1.

Until the ECC has been thoroughly "road-tested" I do not believe there will be many people who actually know how to drive the change-order evaluation procedure. The two fundamental truths common to *all* design and build contracts are that:

- The evaluation of change can only be as sophisticated as the agreed base-line pricing document allows.
- The more the employer seeks to tie down the detail of the pricing document, the more he removes the incentive from the contractor to offer "buildability" and "value engineering" in the design process —particularly if done pre-tender.

Whilst one might be able to value the additional cost of an ECC compensation event, the "Actual Cost" basis will always be the problem area, i.e. how one allows for what might have been, namely the proper value of omitted work, if not self-contained in the priced activity schedule or bills of quantities.

Time is covered by *Core Clause* 3, otherwise referenced as Clauses 30 to 36, which detail what is required of the contractor's programme, including provision for "Float" and "Time Risk allowances"—whatever the distinction between these two might be.

Clause 32 deals with the project manager revising the programme, and Clause 36 covers acceleration—the contractor being required to submit a quotation covering both *cost* and *time*, which the project manager either accepts, so adjusting the agreed completion date and contract price, or rejects. In this event there would appear to be no power for the project manager to insist on acceleration—unless of course the project manager calls in the adjudicator under the disputes procedure of *Core Clause* 9.

Thus in real terms, the only way the completion date gets changed under the ECC contract is through the compensation event procedure and *Core Clause* 6, Clauses 60 to 65. The key provision here is Clause 61.5 and the "strict early warning" requirement, which will preclude "after the event" contractor's claims, which in truth are often no more than damage limitation exercises, trying to find fault with the employer's performance to cover losses incurred by the contractor's own mismanagement.

Certification

Core Clause 5—Payment and Clauses 50 to 54 are a key area, the interim payment provisions in respect of *who* certifies, and *when*, being the same whether under Options A or B. As such, interim payments are due:

- At the end of each "assessment interval";
- At completion of the whole of the works;
- Four weeks after the supervisor has issued the defects certificate; and
- *After* completion of the whole of the works, when either an amount due is corrected (presumably upwards) or when a final payment is made late in respect of interest due.

Under Clause 50.4 the project manager is required to certify, but it is open to the contractor to submit an interim application. In any event the project manager is required to give the contractor details of how the amount certified has been assessed (presumably in writing).

Clause 51 then deals with the mechanics of payment, but, the wording being all in the present tense, it is almost unintelligible. Translated, the procedure is as follows:

- First, establish the first "assessment date";
- The project manager must then certify the first interim valuation within one week of that date; and
- The employer is then due to make that payment within a further two weeks, i.e within three weeks of the assessment date (subject to the parties' right to agree different payment dates);
- If the payment is made late, interest is payable;
- The monthly, or such other regular period as agreed, interim valuation procedure is then, it seems, followed until project completion.

Provision is made at Clause 51.4 for interest being paid in the event of late certification by the project manager and at Clause 51.5 the ECC expressly provides for an interest rate (to be entered in the contract data, Appendix 5) and for compounding of such interest, but annually!

This is a nonsense on two counts and should be amended by the parties in that:

- No bank fails to roll up interest on a much more frequent basis—often daily—so why should a party already being kept out of his money suffer a further loss?
- If one party has been kept out of his money on an interim valuation for a year this is very serious and one must question when the contract determination provisions will kick in.

Presumably the annual compound interest provision envisages adjustments of monies previously paid, as under Clause 95.4 the contractor may terminate if the employer has not paid an amount certified by the project manager within 13 weeks of the date of the certificate.

So we come to the final account—and a big grey area, i.e. the ECC seems to have forgotten to legislate for the necessary procedures, including the legal effect of the final certificate. By implication, therefore, the parties are left to follow the general payment procedure, but I find this profoundly unsatisfactory.

Similarly the whole ECC approach to defects, at *Core Clause* 4, and their signing off is woolly to say the least. Theory is fine, but the real world is somewhat different. Unless express provisions are written into the contract, together with enforceable remedies, non-performing parties will be able to hide because of the inadequacies of the contract drafting. In respect of final account and defects there needs to be much tighter wording and clarity of *who* does *what* and *when*—and *if they don't*, the options the other party has by way of enforcement.

However, the underlying principle of the ECC is that whilst executive authority is vested in the project manager for the purposes of administering the contract, the mere fact that the contractor may challenge any decision of the project manager by calling in the adjudicator should serve as a moderating influence—and hopefully restrict disputes to matters of substance or genuine misunderstanding.

Disputes

Core Clause 9 and Clauses 90 to 92 are a worthy first attempt at writing a disputes regime, based on the concept of a "first-fix" adjudication procedure. However, the ECC provisions are at variance with the Housing Grants, Construction and Regeneration Act 1996, principally in respect of the allowable timescale, and they will no doubt be revised accordingly. Likewise, the "Adjudication Table" setting out what disputes can be referred to an Adjudicator, by whom and when, is concise and helpful but will be consigned to history as under the new legislation there is no restriction on what matters can be raised as a dispute and referred to the adjudicator.

Clause 93 then deals with "Review by Tribunal", i.e. arbitration, presumably—but why not call it arbitration? Otherwise one can envisage all sorts of legal arguments as to how the tribunal is to be set up, how it is to act, and how its decisions are to be enforced! Further, the provision appears to be flawed in that it is unclear as to what the period of limitation is for the dissatisfied party to notify the other party of his intention to refer the adjudicator's decision to the tribunal—is it four weeks as line 5, or "within the time provided by the contract" at line 2 if other than four weeks? Also, some disputes simply cannot be put off until completion, e.g. design issues relevant to ongoing construction.

The ECC therefore, in my opinion, represents a Brave New World and it will be interesting to see how it is rated by those pioneering it on live contracts—albeit the main users appear to be employers offering "partnering" arrangements to "preferred contractors". The cynics amongst us will say such contractors are like muzzled dogs, i.e. bark twice and you are off the list! But only time will tell.

10.6 ICE DESIGN AND CONSTRUCT CONDITIONS OF CONTRACT

General comments

The ICE is a three-party consortium consisting of:

- The Institute of Civil Engineers
- The Association of Consulting Engineers
- The Federation of Civil Engineering Contractors

Further to the issue of the standard ICE Conditions of Contract for Works of Civil Engineering Construction 5th Edition in 1973 the ICE Conditions of Contract Standing Joint Committee recognised the need for two distinct contract approaches covering:

- Contracts based on the employer's design, but with key elements designed by the contractor or specialist subcontractors;
- Contracts where the employer requires the contractor to be responsible for the total design, subject to an outline brief.

The former need has been met by the issue by the ICE Conditions of Contract for Works of Civil Engineering Construction 6th Edition, and the latter need by the issue of the ICE Design and Construct Conditions of Contract.

Accompanying this latter standard form of contract are "Guidance Notes" and the Introduction thereto states specifically:

"The I.C.E. Design and Construct Conditions of Contract depart radically from this concept [Employer's design-based contract settled on an 'admeasure' basis, i.e. measure and value] with the Contractor responsible for all aspects of design and construction including any design provided by or on behalf of the Employer. The Form of Tender provides for payment on a lump sum basis, but other forms of payment may be used."

This role reversal is further emphasised by the continuing advice of the "Guidance Notes" Introduction:

"The Conditions of Contract are therefore not a variant of the 6th Edition, but an entirely new set of conditions based on very different premises. The layout and Clause numbering is similar to those of the 6th Edition but the import of the Clauses is in many cases very different."

The Contents Listing at the front of the standard contract is similarly forthright in that one can readily identify key subjects, for example:

- Clause 1(5) defines "*Cost*" as " . . . all expenditure including design costs properly incurred or to be incurred whether on or off the site and overhead finance and other charges (including loss of interest) properly allocatable thereto but docs not include any allowance for profit."
- Clause 1(6), Communications, which under the contract are required to be "in writing" may be handwritten, typewritten or printed and sent by hand, post, telex, cable or facsimile.

Plain English combined with the reality of today's world—noting the inclusion of the opportunity cost of investment as a "cost"—but why can we not have commas to break up the text and make it easier to read?

On the other hand when one turns to Clause 5(1), Contract Documents one might expect the clause to set out what documents will normally form the Contract Documents, or at least a reference to Articles or an Appendix where such important matters are clarified. With the ICE Design and Contract Conditions the answer lies in the draft "Form of Agreement" enclosed at the back of the standard form.

Similarly one's eyes might light up when one gets down the contents listing to Clause 19(2), Employer's Responsibilities. All one actually finds when turning to the text is not a comprehensive description of such obligations, but a fall-back provision in the event of Clause 31, Facilities for other Contractors, applying—in which case the employer, not the design and build contractor, is responsible for their site safety.

These however are minor criticisms as compared with the general presentation and user-friendliness of this standard form of contract.

Benchmarking of quality

The ICE Design and Construct Conditions do not specifically tell an employer how he should tie down his requirements for *quality* but they certainly spell out sound and, in some respects, unique advice (when compared with other design and build standard forms) as to what he might chose to do.

Firstly, in the "Guidance Notes" under Section 2.2., employer's requirements, the advice is as follows:

"The success of the Contract depends upon a clear understanding by the Contractor of the Employer's Requirements (Clause 1(1)(e)). These are initially set out in the document of that name issued to Tenderers and may be modified prior to the award of the Contract."

The advice continues:

"Their scope may vary from a statement of the overall requirement to a far more detailed description including outline design, specifications, standards of performance, general appearance and other matters."

This simple statement neatly summarises the process behind the production of the employer's requirements document, which is the main plank of the tender enquiry, but in other words it means the employer must have fully considered all aspects of the project brief, including tender strategy and risk evaluation.

Experience shows that ill-considered and improperly defined client briefs, usually put together in limited time, are one of the two main causes for dissatisfaction with design and build. Perhaps therefore greater emphasis should be given in the ICE Design and Build "Guidance Notes" to this essential first step—and to getting it right before going out to tender.

At least the "Guidance Notes" highlight the all-important interim period between tender return and contract award, which represents an opportunity not to be missed. Under JCT 80 lump-sum tendering this is normally a fairly short period—allowing for mathematical checking, second interviews of lowest tenderer, formal employer approval of project commitment and appointment of contractor.

The "Guidance Notes" recognise the opportunity for negotiation and fine-tuning of employer's requirements in the light of the contractor's submission and advise:

"It must be the aim of the parties in discussion, possibly during the tender period and in any event before the award of the contract, so to amend the Employer's Requirements and the Contractor's Submission that they set out an agreed scheme and specification for the Works. In interpreting the final contract the Employer's Requirements take precedence over the Contractor's Submission" (Cl. 5(1)).

Once the project has moved into the post-contract phase, the ICE Design and Build Conditions go further than any other standard form of design and build contract, Clause 8, Contractor's general obligations and Clause 36, Materials and Workmanship providing:

"To the extent required by the Contract the Contractor shall institute a quality assurance system. The Contractor's quality plan and procedures shall be submitted to the Employer's Representative for his prior approval before each design and each construction stage is commenced. (Cl. 8(3)).

" . . . the Contractor shall submit to the Employer's Representative for his approval proposals for checking the design and setting out of the Works and testing the materials and workmanship to ensure that the Contractor's obligations under the contract are met" (Cl. 36(3)).

Finally, in respect of quality control, Clauses 48 and 49 provide clear and workable handover procedures based on "substantial completion"—an improvement, in my opinion, on the JCT concept of "Practical Completion"!

Fitness for purpose

The ICE Design and Construct Conditions specifically stop short of committing the Contractor to "fitness for purpose" liability, in that Clause 8.2(a) provides:

"In carrying out all his design obligations under the contract . . . the Contractor shall exercise all reasonable skill, care and diligence."

As previously discussed this rather defeats the principle of single-point responsibility and puts a greater onus on a disappointed employer being able to plead and prove non-compliance with statute or specification—which may well not be a full materials and workmanship specification on a design and build project.

Rightly the "Guidance Notes" at 6.5 highlight the very real problems of relying on subcontract design consultants taking out professional indemnity insurance let alone maintaining annual run-off. On the other hand, it should not take more than a year for the employer to decide that some aspect of the works are not fit for purpose and to seek redress whilst the contract is still open.

Contract base-line

As previously touched on, the question of what are the base-line drawings is answered under the ICE Design and Construct Conditions by reference to the

Form of Agreement which appears as pages 49 and 50 in the standard contract.

At Item 2:

"The following documents shall be deemed to form and be read and construed as part of this Agreement namely:

    (a) The Conditions of Contract
    (b) The Employer's Requirements
    (c) The Contractor's Submission and the written acceptance thereof
    (d) The following documents . . . "

So, might one ask, what do the ICE recommend as regards a base-line pricing document? How do the ICE envisage interim valuations will be agreed or variations assessed, not to mention how the final account will be agreed?

The ICE-recommended Form of Tender is silent, as are the "Guidance Notes", so it comes down to making proper provision within the employer's requirements document or setting up specific pricing documents which would then need listing under "(d) the following documents" in the Form of Agreement.

This area of base-line pricing is the most difficult area of the design and build procurement process and readers will have noted how the drafting committees of each standard form have danced around the subject. However, the ICE go one better, they have apparently ignored it—both in the express contract provisions and, inexplicably, in the Guidance Notes, where the five short paragraphs commenting on payment provisions are woefully inadequate.

It stands to reason that without a properly defined base-line—whether for drawn information, *quality* or *price*—the question of negotiating and agreeing any changes thereto becomes a lottery—and open season for disputes.

Changes

Guidance Note 7.4(b) recommends that " . . . variations should be limited to matters which are absolutely essential", a sentiment which is easier said than observed. Inevitably the project procurement process will have involved debate as to priorities in space usage, affordability of the built volume and the level of services and finishings therein. Usually compromises will have been reached before tenders are obtained, so it follows that during the build-out process there will be strong temptation to ask for more.

Typically this manifests itself in the course of approving the design and build contractor's design development proposals, by employer or his representative suggesting alternatives to the contractor. It then just needs a commercially aware contractor's QS and before you know it there is a "Confirmation of Verbal Instruction" or some such contractor's standard pro forma on your desk—which, if you do not bounce it immediately, qualifies as a variation under Clause 52(2), via the default provision of Clause 2.5(6), Confirmation of Oral Instructions.

Otherwise the ICE Design and Construct Conditions provide at Clause 51(1) that:

"The Employer's Representative shall have power after consultation with the Contractor's Representative to vary the Employer's Requirements. Such variations may include additions and/or omissions and may be ordered at any time up to the end of the Defects Correction Period for the whole of the Works."

Quite what the purpose of the provision for "consultation" is, apart from common courtesy, can only be guessed at—but what if the Contractor's Representative does not want to play? Can the employer's representative issue the instruction varying the employer's requirements regardless?

Then we come to the key area of "fair and reasonable" valuation of variations. The ICE Design and Build Conditions start from the assumption that such can be amicably agreed and that, whenever possible, this will be done before the work is put in hand—Clause 52(1). Unfortunately the real world, particularly that of the construction industry, rarely works by consensus when money and contractual risks are involved.

So the ICE Design and Build Conditions, Clause 52(2) provides the following mechanism for the administration of variations:

- Contractor submits, as soon as possible:

  (a) (i) "His quotation for any extra or substituted works necessitated by the variation having due regard to any rates or prices included in the Contract, and . . .
     (ii) his estimate of any delay occasioned thereby, and . . .
     (iii) his estimate of the cost of any delay".

- The employer's representative then has 14 days to:

  (b) (i) "Accept the same or . . .
     (ii) negotiate with the Contractor thereon".

If agreement is reached all well and good, but if not:

" . . . the Employer's Representative shall notify the Contractor of what in his opinion is a fair and reasonable valuation and thereafter shall make such interim valuations for the purposes of Clause 60 as may be appropriate."

The interesting reference is in (a)(i)—previously the "Guidance Notes" said the ICE Design and Build Contract was primarily a lump sum as opposed to an admeasure contract, yet here we have a suggestion that there might be schedules of rates or bills of quantities. The real problem is that whilst one might be able to reasonably agree how to value the *cost* of the newly instructed work—reverting to recorded costs and dayworks if necessary—quite how one splits out from a lump sum the hidden cost of omitted work will inevitably test the objectivity of the two parties' representatives.

Equally the proper adjustment of *time* does depend upon a recognisable base-line, usually a construction programme showing a declared critical path against which any instructed variation to the employer's requirements can be factually plotted.

Thus the employer's representative retains ultimate authority, but this is subject to the unhappy contractor's right to invoke the settlement of disputes procedure, as provided by the ICE Design and Construct Conditions, Clause 66.

Delays

By not providing for compulsory pre-costing of instructed variations, the ICE Design and Construct Conditions approach to adjusting the twin goalposts of *cost* and *time* are in my view passive and ineffective—leaving too much to chance that the parties might agree or, more likely, not agree. This makes for uncertainty, which is not in the interests of either the employer or the contractor, being the catalyst for disputes and wasted costs in pursuing perceived rights.

Condition 44(1), Extension of Time for Completion, provides that:

"Should the Contractor consider that:

    (a) any variations ordered under Clause 51(1) or . . .

    (b) any cause of delay referred to in these Conditions or . . .

    (c) exceptional adverse weather conditions or . . .

    (d) other special circumstances of any kind whatsoever which may occur

be such as to entitle him to an extension of time for the substantial completion of the Works or any Section thereof he shall within 28 days after the cause of any delay has arisen or as soon thereafter as is reasonable deliver to the Employer's Representative full and detailed particulars in justification of the period of extension claimed in order that the claim may be investigated at the time."

So at least there is a cut-off under the ICE Design and Construct Conditions, which will effectively shut out retrospective claims—usually hurriedly invented when contractors perceive liquidated and ascertained damages to be a real threat.

Condition 44(2) then provides for "assessment of delay" by the employer's representative but fails to impose any time limit for such assessment, whether deliberately or by oversight, but this is restricted to " . . . the delay [if any] that has been suffered . . . ", i.e. a historic assessment which by definition cannot be more than 28 days, given the cut-off under Condition 44(1).

The more difficult area of judgement falls under Condition 33(3), the ICE Design and Construct Conditions opting for the granting of an interim extension, followed by a final review under Condition 44(4). Such final review must, according to Condition 44(5), be made and published by the employer's representative within 14 days of the issue of the certificate of substantial completion of the works.

The ICE Design and Construct Conditions then provide at 46(1) for the employer's representative ordering acceleration at no additional cost, if in his opinion the completion date is not achievable, or, if he wishes to bring the completion date forward he may so instruct under Condition 46(3), but only by agreement with the contractor as to the costs involved.

Inevitably, there may well be disagreement as to the employer's representative's considered decisions in respect of *time*, so the contractor's protection under the ICE Design and Construct Conditions lies at Condition 66(1), Settlement of Disputes.

Certification

The interim payment mechanics of the ICE Design and Construct Conditions are provided at 60(1) and (2). Briefly summarised these provide that:

- The contractor submits to the employer's representative at such time and in such form as the contract prescribes statements showing:

    (a) the amounts which in the contractor's opinion are due under the contract

    and where appropriate showing separately

    (b) Contingency and prime cost expenditure;
    (c) Goods and materials on-site but unfixed;
    (d) Goods and materials off-site;
    (e) All other matters, e.g. preliminaries, variations, etc.

- The employer's representative must then certify what he considers to be properly payable, without having to state any value or reasons for values requested but not certified and the employer must then pay—all within 28 days.

The above provisions must assume the base-line pricing document being in some detail, whether agreed stage payments with elements and sub-elements priced by lump sums, or possibly a short form bill of quantities. Otherwise the scope for the contractor over-claiming and the employer's representative under-certifying will invite disputes and a breakdown in relationships at site level.

Any standard contract which not only gives unilateral power to the employer through his agent effectively to decide how much he wishes to pay and when, but also fails to impose any obligation on the employer's agent to explain and discuss his decision with the contractor is, in my view, asking for trouble.

The final account procedure under the ICE Design and Construct Conditions, Condition 60(4) repeats the interim payment requirement, i.e.:

- The contractor submits not later than three months after the date of the defects correction certificate.
- The employer's representative issues a final certificate at his sole discretion within a further three months, or three months after receipt of " . . . all information reasonably required for its verification".
- The employer then pays within a further 28 days.

At Condition 60(7) the ICE Design and Construct Contract scores full marks, where all others fail. Not only is there provision for interest on late payments but it also provides for rolling-up of interest every month and prescribes a rate of 2 per cent above base lending rate.

Further, at Condition 60(8) there is provision for the employer's representative disallowing for payment purposes any workmanship or materials with which he is not entirely satisfied, without contractually being required to so identify. Clearly, however, it would be unreasonable to stop monies without giving the contractor an explanation, together with a requirement to make good the defective work.

Finally Condition 60(10) opens up an interesting scenario. It provides as a contractual requirement that, should the employer not honour any employer's representative's payment certificate in full, he must give full details of any reduction to the contractor. Other than the deduction of liquidated and ascertained damages under Clause 47 it is difficult to envisage circumstances under the contract which would allow the employer to do this.

Thus the employer's representative is the undisputed judge of *quality*, the ICE Design and Construct handover procedure being:

48(1): The contractor notifies the employer's representative when he considers he has achieved substantial completion, and gives an undertaking to complete any outstanding work.

48(2): The employer's representative then has 21 days to either:

(a) Accept the contractor's submission and issue a Certificate of Substantial Completion, or

(b) Identify in writing to the contractor all the work which needs to be completed or made good before such certificate will be issued.

49(1): The contractor completes all outstanding work, preferably at agreed times.

(2) The contractor delivers up each section and the whole of the works at the end of the relevant defects correction period.

61 (1) The employer's representative inspects and, if satisfied, issues a Defects Correction Certificate.

Finally in respect of certification, it is interesting to note that the ICE Design and Construct Guidance Notes at 7.6 spell out the responsibilities of the employer's representative, particularly when dealing with payments, with issues of substantial completion and with finally signing-off the works. This reads:

"The Employer's Representative has certain decision making functions under the Contract. These functions and the codifying functions listed under Section 7.5(4) and 7.5(5) of these Guidance Notes require to be exercised with the same fairness and impartiality which is identified with decisions of the Engineer under the other I.C.E. Conditions of Contract."

Whether this recommendation is intended to redress the obvious imbalance of power vested in the employer's representative, it only appears in the Guidance Note and is not an express term of the contract.

Disputes

The ICE Design and Construct Conditions have a slightly different approach to the resolution of disputes, albeit they will need to be amended to provide for adjudication as required under the Housing Grants, Construction and Regeneration Act 1996. The procedure can be summarised thus:

- The employer and the contractor can dispute virtually any matter if they so wish at any time. There is no cut-off or provision of a period following the issue of the final certificate after which the final certificate becomes final and binding (Cl. 66(1)).
- The unhappy party must first exhaust all steps specifically provided by the condition covering the relevant issue and only then may he serve a Notice of Dispute on the other party (Cl. 66(2)).
- There shall then be a one-calendar-month negotiation/cooling-off period before the parties may proceed to the next stage.
- After the said one month, if no settlement has been achieved " . . . the Notice of Dispute is referred [by whom?] to conciliation in accordance with the Institution of Civil Engineers' Conciliation Procedure (1988) or any modification thereto being in force at the date of such referral".
- "The Conciliator shall make his recommendation in writing and give notice of the same to the Employer and the Contractor within three calendar months of the service of the Notice of Dispute" (Cl. 66(3)).
- In the meantime the parties are required to proceed with the works, including any recommendations of the conciliator (Cl. 66(4)).
- If either party is still dissatisfied they then have three calendar months from the conciliator's recommendations to serve Notice of Arbitration (Cl. 66.5(a)), or,
- Should the conciliator fail to give his recommendations within three calendar months of reference to conciliation, and the passage of a further three calendar months, then either the employer or contractor may serve a Notice of Arbitration (Cl. 66.5(b)).
- If the parties fail to concur in the naming of the arbitrator, an arbitrator will be appointed by the President or Vice-President of the Institution of Civil Engineers (Cl. 66(6)).
- Provision is made for the arbitration procedure (Cl. 66(7)).

Interestingly Clause 66(8) then provides that:

- The employer's representative may be called as a witness before either the conciliator or the arbitrator, and

- All submissions to the conciliator are without prejudice and the conciliator may not be called as a witness in any arbitration or subsequent legal proceedings.

10.7   FIDIC CONDITIONS OF CONTRACT FOR DESIGN-BUILD AND TURNKEY

General comments

FIDIC is the one institution in our industry that seems to be a permanent fixture and goes back to the days when in construction terms Great Britain and France led the world, when the USA were far from United, being busy building ironclads and military railroads to sort out their internal squabble about slavery!

The Fédération Internationale Des Ingénieurs-Conseils (FIDIC) therefore led the world in establishing the contracting ground rules as Great Britain and France opened up colonies and other territories in their search for raw materials—minerals and food stuffs. Their well proven standard form of contract has been progressively updated and fine-tuned to meet the demands of today's world.

As foreign countries won independence inevitably many situations required European construction teams not only to provide design, as the locals had no design expertise, but also to operate the facility and to train local staff before finally handing over. As a result, the concept of "turnkey" contracting emerged where the foreign employer organisation would take little interest in the project until all was built, commissioned and demonstrated, i.e. the moment of the figurative handing-over of the key to the front door, with or without the option of further hand-holding by way of operative training.

This led to many standard FIDIC contracts being adapted to incorporate design as part of the contractor's obligation, but now this scenario has been covered by a bespoke Design-Build and Turnkey contract FIDIC version. Known as the "Orange Book", this was first published in 1995 and is a welcome addition, having been produced by a task force of engineers and lawyers drawn from Germany, UK, Canada, USA and Sweden.

In the Foreword to the Orange Book the point is made that such a standard form facilitates international competitive tendering, but also that with minor modifications the conditions are suitable for use on domestic contracts. As the Foreword also points out:

"There are no universally-accepted definitions of the terms 'design-build' and 'turnkey', except that both involve the contractor's total liability for design. For the Employer, such single-point responsibility may be advantageous, but the benefits may be offset by having less control over the design process and more difficulty in imposing varied requirements."

Readers who are not familiar with international contracting should also appreciate that there are often two employers on an international project, i.e.

the project employer being the local organisation or end user, often a government organisation of whatever country requiring the facility and the contract employer, who not infrequently will be a UK Government or UK commercial organisation for whom the construction contract forms but part of a larger trade deal.

Readers should also appreciate that once one physically gets to an overseas project location it can be a lonely place. If the location happens to be where the locals profess to be "holier than thou" on all their flags you will be playing a distinctly "away" fixture with, usually, three sets of rules:

- The contract rules;
- The official local rules—which can be open to wide interpretation by local individuals with the right family or political connections;
- The unofficial local rules—which may well involve "fix-it" men, off-shore bank accounts, crates of consumables, or local trade-offs.

For example, on one large bank headquarters project, we needed the permanent water and electricity on before the extensive planting (including semi-mature trees) could commence. The local Mayor decided we could only have water and electricity when all was planted *unless* we built a surfaced football field half a mile away complete with surrounding wall!

I duly reported this "deal" option to the employer and the local Mayor was promptly relieved of his duties, as one of the bank directors was the son of the Ruler. We had water and electricity within 48 hours and the local kids continued to play football in the street.

Further, it often happens that there is a lead UK contractor on an overseas project, but that the specialist installations are the "*raison d'être*" and represent the larger project value, e.g. munitions manufacturing complexes. If so, the specialist subcontractors must go through the main civil works contractor for visas etc. and will hold a FIDIC-contract where the UK lead contractor is named as the "Employer" and the specialist subcontractors are each named as the "Contractor" in the respective documentation.

It can also happen that the head contract or protocol agreement between the two foreign countries has terms and conditions, and therefore risks, which are not entirely back-to-back with the FIDIC-based contract as entered into between the UK lead contractor (the employer) and the specialist subcontractors (contractors). It is therefore important always to know at what level of the project one is dealing when using the legal terms "employer" and "contractor" and to know where the real risks lie.

These niceties of overseas contracts are discreetly referred to by FIDIC Clause 2.3:

"The Employer shall, at the request and cost of the Contractor, assist him in applying for permits, licences or approvals, which are required for any part of the Works, for delivery (including clearance through customs) of Plant, Materials and Contractor's Equipment, and for completion of the Works. Such requests may also include requests for the Employer's assistance in applying for any necessary government consent to the export of Contractor's Equipment when it is removed from site."

In my experience "local conditions" or "co-location" make for sensible behaviour *between* the UK parties to an overseas contract from main contractor downwards—one simply tends to pull together in the common cause, helping one another, and swapping stories about the locals in the bar or site club after work. Even if subcontracts are placed with local firms, one needs them to perform, so one looks after them as best one can rather than throw the contractual book at them.

Perhaps the UK construction industry is now just starting to understand that this alternative approach can work—if only all main parties to a project can see the mutual benefit of talking informally and avoiding adversarial attitudes.

Anyhow, enough "soap-boxing" and on with the action as to the key provisions of the FIDIC Design-Build and Turnkey contract.

Benchmarking of quality

In international contracting there is usually a greater emphasis on the civil engineering elements and function of the project than there is on architecture and *quality* of finishings. Nevertheless, it is particularly important to pay attention to local conditions such as geology, climate, availability of materials and labour, etc., and to write a bespoke specification for each project, supported by drawings as necessary.

If contemplating a design and build procurement route, the lead UK contractor or agency is likely to have:

- Identified the foreign client's expressed needs;
- Blended these with perceived needs, particularly in defence-industry related projects;
- Secured project funding, in principle if not in fact, and now be looking for UK contractors, acting singly or in consortia, to take *all* the risks of project development and execution.

On this basis, the lead contractor's "Specification" document is likely to be drawn in the widest and most non-specific terms as to how the foreign organisation's identified and perceived needs are to be met.

At this stage the project is likely to be broken up into its principal constituent parts, e.g. civil works including infrastructure installations, buildings, etc., and major subcontract partners sought, each of whom will take the design risk in their particular field. As part of this process, schematic or more detailed drawings will be prepared by the preferred UK subcontractors and full performance specifications drafted which in turn will allow the lead UK contractor to take the proposal forward and hopefully sign the second-stage head contract or protocol agreement with the foreign client.

Inevitably the temptation will be two-fold:

- Better the devil you know than the devil you don't know;
- Why not promote UK industry?

so if you have the task of drafting the necessary "Specification" you naturally think "British" and lace the material and workmanship provisions with BS references. But these may be wholly inapplicable, in that:

- They are based on different geological or climatic criteria;
- They ignore the foreign client's desire to promote his own established or emerging industries, which are often protected by import restrictions;
- They ignore the valuable benefit of the world market for material purchasing.

The usual answer is to provide other national standards references, e.g American standards, as an alternative to British Standards—and leave the site engineer to work out whether the load of structural steel just delivered to his remote site actually complies with either, let alone where it might have originally been manufactured!

So one cannot be too particular in drafting performance specifications for overseas projects and should perhaps concentrate on defining unacceptable standards in general terms in respect of materials and workmanship, leaving the site engineer as the final arbiter. However, where one can take a tight line is in requirements for documentation control and production, especially shipping documentation. This will be necessary not only for matters of *quality* control, but also insurances and payment.

Not surprisingly therefore the FIDIC Design-Build and Turnkey contract makes no provision except at Clause 5.4, Technical Standards and Regulations, for how an employer should express his requirements as to the scope of design required prior to tender. The only restraint, as found at Clause 5.4, is that the developed design, and execution thereof, shall be in accordance with the particular country's national specifications, standards and law—assuming such are published; but this should not preclude the setting of a higher standard, if deemed appropriate.

Fitness for purpose

Clause 5 of the FIDIC Design-Build and Turnkey contract deals with design, the Foreword having spelt out the basic concept of single-point responsibility, i.e. design is the contractor's risk.

Clause 5.1 takes no prisoners in providing:

- "The Contractor shall carry out, and be responsible for, the design of the Works . . . "
- "Design shall be prepared by qualified designers who are engineers or other professionals who comply with the criteria (if any) stated in the Employer's Requirements . . . "

- "For each part of the Works, the prior consent of the Employer's Representative shall be obtained to the designer and design subcontractor, if they are not named as such in the Contract . . . "
- "The Contractor holds himself, his designers and design subcontractors as having the experience and capability necessary for the design . . . "
- "The Contractor undertakes that the designers shall be available to attend discussions with the Employer's Representative at all reasonable times during the Contract Period."

Clause 5.2 continues in similar vein in laying down the ground rules for design development and approval thereof:

- "The Contractor shall prepare Construction Documents in sufficient detail to satisfy all regulatory approvals, to provide suppliers and construction personnel sufficient instruction to execute the Works, and to describe the operation of the completed Works . . . "
- "The Employer's Representative shall have the right to review and inspect the preparation of Contract Documents, wherever they are being prepared . . . "
- "Each of the Construction Documents shall, when considered ready for use, be submitted to the Employer's Representative for pre-construction review."

The employer's representative then has to review such submissions within 21 days (defined as calendar days at Clause 1.1.3.6). Should he not approve the drawings for construction but require further amendments, the process is repeated again with a potential further 21-day approval period.

In the meantime no work as covered by the contractor's design development submission can officially commence and the provisions then have a sting in the tail:

- "Errors, omissions, ambiguities, inconsistencies, inadequacies and other defects shall be rectified by the Contractor at his cost."

Contract base-line

Clause 1.5 of the FIDIC Design-Build and Turnkey contract clearly anticipates a period of negotiation between tender and formal contract signature in that it provides:

"Either party shall, if requested by the other party, execute a Contract Agreement, in the form annexed with such modifications as may be necessary to record the Contract."

Assuming a formal Contract Agreement is then executed, Clause 1.6 provides for the following "Priority of Documents":

(a) The Contract Agreement
(b) The Letter of Acceptance
(c) The Employer's Requirements
(d) The Tender
(e) The Conditions of Contract, Part II
(f) The Conditions of Contract, Part I
(g) The Schedules; and

(h) The Contractor's Proposal

The foregoing list is interesting for a number of reasons:

1. Conditions of Contract, Part II, the Particulars relevant to the project, take precedence over Part I, the General Conditions in the event of conflicting provisions.
2. The Employer's Requirements take precedence over the Conditions of Contract, e.g. if the Employer's Requirements call for "Fitness for Purpose" but the Conditions call for the lesser duty of "Reasonable skill and care" under the FIDIC Design-Build and Turnkey contract the higher duty will prevail, whereas under JCT 81 the Contract Conditions and the lower duty would govern the project.
3. Normally it is considered that on formal contract signature any Letter of Acceptance, Letter of Intent, exchange of correspondence, etc., is superseded—but not so under FIDIC Design-Build and Turnkey, where the Letter of Acceptance remains a contract document.
4. However item (g) "The Schedules"—which will set the ground rules for the administration of the project—is likely to be where most attention should be focused, as without specific provisions for the mechanics of payment and currency exchange risk, established by way of "engagement" between the parties, any subsequent "marriage" is likely to be a stormy one.

Helpfully FIDIC have published, as part of the Orange Book:

- Part II Guidance for the Preparation of Conditions of Particular Application, and
- Forms of Tender and Agreement

The guidance document suggests various example sub-clauses under each of the principal clauses of Part I, General Conditions. Of these, Clause 13, Contract Price and Payment, is essential reading and provides, *inter alia*:

- "When writing Part II [the bespoke contract provisions] consideration should be given to the amount and timing of payment(s) to the Contractor . . . "
- "A positive cashflow is clearly of benefit to the Contractor, and tenderers will take account of the interim payment procedures when preparing their tenders . . . "
- "Normally, this type of contract is based on a lump sum price, with little or no remeasurement; the Contractor then takes the risk of changes in cost arising from his design . . . "
- "The lump sum price may consist of two or more amounts, quoted in the currencies of payment (which may, but need not, include Local Currency . . ."
- "In order to value Variations, the tenderer can be required to submit a detailed breakdown of the Contract price, including quantities, unit rates and other pricing information; such information can also be used for the Interim Payment Certificates . . . "

The Guidance Notes then make a rather strange statement in the form of a warning:

> • "However, that information may not have been competitively priced; when the tender documents are being prepared, the Employer must decide whether he is prepared to be bound by such information. If not, he should have ensured that the Employer's Representative has the necessary expertise to value such Variations which may be required."

I can just imagine the number of committee hours involved in coming up with this "fudge"—but "fudge" it is, which is a shame given the general clarity of thinking, and drafting, of this FIDIC contract. Unless one imposes a proper cost-control regime as part of the tender enquiry one is asking for trouble. The apparent suggestion that one asks the successful tenderer for cost build-ups *after* the event, and without being sure how they relate to the total accepted contract price, leaves far too much to chance and trust, in my experience.

When sitting in a London office setting up the basic rules for an overseas project costing and payment procedures, I will focus on the following aspects which generally do *not* apply to UK contracts:

• Who will take the currency exchange risk, how might this change during the course of the project and how can one limit such risk by managing payment terms?

• How will one be able to simply assess value of work done given the remoteness of project location and limited staff resources? For instance, one "project" I controlled from London involved 32 sites all over Saudi Arabia—requiring just one co-ordinated interim payment certificate each month.

• Lastly, but probably the most important of all, how will one be able to demonstrate to the foreign client that the claimed amount of work has actually been carried out and that he should therefore release the next tranche of money to the UK contracting team? Again, I can speak from experience of having had to deal with the Ministry of Finance in Tehran and the Ministry of Housing in Baghdad on different projects, at different times and holding different passports!

I could write a separate book on this fascinating subject, but one must *not* assume the foreign client:

• Will be technically represented, or
• Will necessarily have a payment authorisation procedure and duly experienced people in place to administer it, or
• Will necessarily wish to pay!

Essentially the trick when setting up foreign contracts, especially in the more volatile parts of the world, is to write formal advance payment provisions backed by a UK bond, and then to gear payments to monthly "finger in the wind" assessments of work done, i.e. simple visual inspection of the works by defined parts (e.g. individual buildings), by defined elements (e.g. foundations, structure, finishings, services) plus infrastructure works and defined key plant which is in transit, supported by insurance and shipping documentation.

The currency exchange risk is best neutralised by identification of *where* resources and materials will be primarily purchased at each stage of the project. For example the FIDIC Design-Build and Turnkey contract Conditions, Clause 13.1(6) provides: "The Contract Price shall not be adjusted for changes in the cost of labour, material or other matters."

This is fine: one can price the direct costs of labour and material and reasonably estimate price rises over the duration of the project. However, if one is being paid entirely in US dollars or in sterling one also has to purchase local currency to pay local labour.

As local currencies can be volatile against world money markets, especially in developing countries, this risk is best managed by organising the contract such that in broad terms local costs incurred are paid out in local currency, and offshore costs are paid for in a recognised international currency, e.g. US dollars. In this way, the foreign client pays the UK lead contractor in two currencies each month in predetermined ratios. For example, initial project costs, i.e. expat staff salaries paid in sterling direct to UK or offshore banks, major material purchases and UK subcontract costs might represent 80 per cent of cost exposure over the first six months, leaving 20 per cent local setting-up, UK expatriate staff local pay and minimal site labour costs. As the project reaches finalisation, the UK and local cost balance may be exactly the other way round, i.e. 20 per cent UK against 80 per cent falling to be paid in local currency.

So one writes a simple provision that all monthly payments will in the first instance be calculated in whichever single currency the contract pricing document has adopted, then, according to the month number, a predetermined proportion is paid in the second currency. For further information, the Guidance Notes on Clause 13.15 are first-rate, as is the subsequent advice on "project financing agreements".

Generally, however, the idea is that the whole project will remain forward-funded via the advance payment provision. This is progressively repaid throughout the project, but is an essential commercial safeguard as against the UK lead contractor having to pull out due to war, local politics, or simple non-payment. In this event the UK subcontractors who have closely supported the UK lead contractor will not expect to be cut off at the knees, but to be looked after by way of having their abortive and contract determination costs paid, in whole or in part, by the UK lead contractor out of what remains of the advance payment fund.

Thus FIDIC Design-Build and Turnkey contract, Clause 13.2 provides for "advance payments" and for their repayment, whether conventional monthly interim payments (Cl. 13.3) or schedule payments (Cl. 13.4) are to apply.

Changes

The change order procedure under the FIDIC Design-Build and Turnkey contract is covered by Clause 14 and provides for two categories of variations:

- Variations instructed by the employer's representative under Clause 14.1.;
- Contractor-proposed variations of a "value engineering" nature under Clause 14.2 identifying a potential reduction in "the cost of constructing, maintaining or operating the Works, or improving the efficiency or value to the Employer of the completed Works, or otherwise be of benefit to the Employer".
- The employer's representative may request a submission from the contractor under Clause 14.3 setting out:

    (a) A description of the work, together with an assessment of the design and time for carrying out the work;
    (b) Detailed implications of the proposed work on the construction programme;
    (c) The proposed adjustment to the contract price, a revised time for completion (if applicable) and any other required modifications to the contract.

The employer's representative may of course accept such proposals, but is otherwise bound by Clause 3.5:

"When the Employer's Representative is required to determine value, cost or extension of time he shall consult with the Contractor in an endeavour to reach agreement. If agreement is not achieved, the Employer's Representative shall determine the matter fairly, reasonably and in accordance with the Contract."

Again, fine sentiments, but if there is no effective base-line pricing documentation one wonders just how any employer's representative will be able to do this *and* satisfy the contractor.

As regards prime cost and provisional sums the FIDIC Design-Build and Turnkey contract only recognises the latter, which might be applicable when the employer needs to reserve certain matters for specific selection or to provide for unknowns such as ground conditions. In each case the provisional sum should be defined as to scope and, when expenditure is authorised, priced at net cost plus pre-tendered on-costing to cover all other costs, charges and profit.

Delays

The FIDIC Design-Build and Turnkey contract deals with "extension of time for completion" at Clause 8.3. The provisions are simple and fair, requiring the contractor to notify any possible claim within 28 days of the start of the event giving rise to the delay.

The contract specifically requires the contractor to retain relevant contemporaneous records, either on site or at an agreed off-site location, and further provides for an "open book" inspection regime.

Of particular interest is the provision at Clause 8.4 which reads:

"If the following conditions apply, namely:

(a) The Contractor has diligently followed the procedures laid down by the relevant legally constituted public authorities in the Country . . .
(b) Such authorities delay, impede or prevent the Contractor, and . . .
(c) The resulting delay to the Works was not (by the Base Date) foreseeable by an experienced contractor,

then such delay will be considered as a cause for delay giving an entitlement to extension of time under Sub-Clause 8.3."

Again the employer's representative is required to have the wisdom of Solomon under Clause 8.3, being required to "agree or determine either prospectively or retrospectively such extension of Time for Completion as may be due", subject to a further provision that "he shall review his previous determinations and may revise but shall not decrease, the total extension of time".

One further safety valve provided by the FIDIC Design-Build and Turnkey contract is the option at Clause 8.7 for the employer's representative to instruct the contractor to suspend the works. The FIDIC provisions allow the employer to terminate under Clause 2.4, but *there is no equivalent provision for termination by the contractor.* This suspension option would therefore kick in should the foreign client fail to honour his payment allegations—this is assuming the employer's representative is independent of such foreign client and in appropriate circumstances will give the necessary instruction under Clause 8.7.

In the event of suspension and subsequent recommencement Clause 8.8 provides for the proper reimbursement of the contractor.

Certification

In discussing the contract base-line and the need for pricing documentation I have touched on the interim payment requirements of Clause 13. Assuming monthly interim payments are provided for by the contract Sub-Clause 13.3 requires the contractor to submit six copies to the employer's representative " . . . after the end of each month, in a form approved by the Employer's Representative, . . . together with supporting documents which shall include the detailed report on the progress during the month in accordance with sub-clause 4.15".

The linkage of progress reports to payment entitlement is intriguing, but why six copies are necessary, particularly when back-up shipping documentation might be involved, is not clear, as it is surely excessive. The employer then has 56 days " . . . from the date on which the Employer's Representative received the Contractor's statement and supporting documents . . . " to effect payment under Clause 13.7.

However, quite what the position is as regards the period for payment by the employer if Clause 13.4, Schedule of Payments, applies is not clear. Presumably the FIDIC contract assumes that each scheduled payment will be defined by progress on site, as envisaged by Sub-Clause 13.4(c), requiring the employer's representative to so certify—which will then trigger the 56 days for payment by the employer under Clause 13.7.

Disputes

The FIDIC Design-Build and Turnkey contract closes with Clause 20, Claims, Disputes and Arbitration.

Sub-Clauses 20.3 and 20.4 focus on the appointment of an Adjudication Board in the event of a claim not being settled by the employer's representative to the contractor's satisfaction. These two Sub-Clauses are relatively lengthy, unlike the general structure of the FIDIC contract provisions, but on large overseas projects it may well be necessary to have a three-person Board and a detailed set of procedural rules.

As the Guidance notes in Part II make clear, in commenting on Sub-Clause 20.3, the agreement of suitably qualified independent Board members is best done pre-contract, but not at the cost of delaying the project or falling out as to who should be appointed! However, Sub-Clause 20.5, Amicable Settlement, is the obvious preferred course of action and alone of all the standard forms FIDIC give it headline billing, by virtue of an express contract provision.

As the Guidance in Part II points out, "amicable settlement procedures depend for their success on confidentiality and on agreement of the procedure", but if this fails Sub-Clause 20.4 helpfully provides a first stage adjudication procedure as follows:

- "If a dispute arises . . . the dispute shall initially be referred in writing to the Dispute Adjudication Board for its decision . . . "
- "No later than the fifty-sixth day after the day on which it received such reference, the Dispute Adjudication Board, acting as a panel of expert(s) and not as arbitrator(s), shall give notice of its decision to the parties . . . "
- " . . . the Contractor and the Employer shall give effect forthwith to every decision of the Dispute Adjudication Board, unless and until the same shall be revised, as hereinafter provided, in an amicable settlement or an arbitral award."

Thus the concept of an agreed *ad hoc* "amicable settlement" procedure comes *after* the initial adjudication hearing but before the *final* formal arbitral hearing. The FIDIC Clause 20.4 provisions continue:

- "If either party is dissatisfied with the Dispute Adjudication Board's decision, then either party, on or before the twenty-eighth day after the day on which it received notice of such decision, may notify the other party of its dissatisfaction . . . ."

- "If the Dispute Adjudication Board has given notice of its decisions as to a matter in dispute to the Employer and the Contractor and no notice of dissatisfaction has been given by either party on or before the twenty-eighth day after the day on which the parties received the Dispute Adjudication Board's decision, then the Dispute Adjudication Board's decision shall become final and binding upon the Employer and the Contractor."

So, simply summarised, unless either party wishes to keep a dispute alive they *must* take a further formal step within 28 days of the Dispute Adjudication Board's decision.

Sub-Clause 20.5, Amicable Settlement, then provides an effective cooling-off period for a *further* 56 days before arbitration may be commenced. Whether "commencement" in this context means the serving of Notice of Arbitration by one party on the other, or the start of the actual hearing is not stated, but must in the spirit of Sub-Clause 20.5 mean the former.

There then follows Sub-Clause 20.6, Arbitration, which provides for international arbitration and sweep-up provisions in the event of one party not complying with any uncontested decision of the Dispute Adjudication Board under Sub-Clause 20.7.

So there you have it—a very different, thought-provoking standard form of design and build contract, which, whilst specifically packaged for export, nevertheless has some very persuasive and simply worded clauses which could, in my opinion, be usefully re-imported into further drafts of the standard UK design and build contracts as discussed earlier in this book.

JCT 81 WCD SUPPLEMENTARY PROVISIONS

S1 Adjudication

Article 5 and Clause 39 (*settlement of disputes—Arbitration*) shall have effect as modified by the following provisions:

Dispute or Difference—Adjudication Matters—Reference to Adjudicator

S1.1 If a dispute or difference has arisen between the Employer and the Contractor prior to Practical Completion or alleged Practical Completion of the Works or termination or alleged Termination of the Contractor's employment under this Contract or abandonment of the Works on any one or more of the matters set out in S1.2 (Adjudication Matters) that dispute or difference shall not be referred to arbitration under Clause 39 but shall, as provided in S1.3, be referred to the Adjudicator named in Appendix 1 or appointed under S1.5.

The Adjudication Matters

S1.2 The Adjudication Matters are:
- .1 any adjustment or alteration to the Contract Sum; or
- .2 whether the Works are being executed in accordance with the Conditions and the Supplementary Provisions; or
- .3 whether or not the issue of an instruction is empowered by the Conditions; or
- .4 whether either party has withheld or delayed a consent or statement or agreement where such consent or statement or agreement is not to be unreasonably withheld or delayed; or
- .5 on any dispute or difference under Clause 4.1.1 (*a Change affecting obligations or restrictions imposed by the Employer*) in regard to a reasonable objection by the Contractor, or under Clause 8.4 (*Powers of Employer—work not in accordance with the Contract*), or under Clause 8.5 (*Powers of Employer—non-compliance with Clause 8.1.3*), or under Clause 17.1 (*partial possession*) or under Clause 23.3.2 (*early use or occupation by Employer*) in regard to a withholding of consent by the Contractor, or under Clause 18.2.1 (*sub-letting of work*) or under 18.2.3 (*sub-letting of design*) in regard to a withholding of consent by the Employer, Clause 25 (*extension of time*) or under the Supplementary Provisions and in particular under S2.2, S3.2, S4.2.2, S4.3.1, S4.4.1, S5.2, S6.2, S6.6, S7.4.

Decision of Adjudicator

S1.3 .1 Either the Employer or the Contractor may give to the other written notice that a dispute or difference as referred to in S1.1 has arisen. Not later than 14 days after the date of that notice the parties shall in statements to the Adjudicator set out the matters in dispute on which the decision(s) of the Adjudicator are required.

.2 Within 14 days (or within such time as the parties may agree) of receipt of the statement by the parties the Adjudicator shall inform the parties when he expects his decision(s) will be given and may require from either party such further information or documents as he reasonably requires to enable him to reach his decision(s). If either party fails to comply with any such requirement such failure shall not invalidate the Adjudicator's decision. In giving his decision the Adjudicator shall be deemed to be acting as expert and not as arbitrator.

.3 Subject to S1.4 the decision(s) of the Adjudicator shall be deemed to be a provision of this Contract (an Adjudicator Provision) and such Adjudicated Provisions shall be final and binding on the parties unless referred to arbitration as provided in S1.3.4. or S1.4. If there is any conflict between an adjudicated provision and any other provision of this contract the Adjudicated Provision shall prevail.

.4 If within 14 days of receipt of the decision(s) of the Adjudicator either party informs the other in writing that any such decision is not acceptable then such decision shall nevertheless remain an Adjudicated Provision of this Contract; but the dispute or difference on which the decision of the Adjudicator is not acceptable is referred to arbitration in accordance with Article 5 and Clause 39, but such arbitration shall not be opened until after Practical Completion of the Works or alleged Practical Completion of the Works or termination or alleged termination of the Contractor's employment under this Contract or abandonment of the Works. Provided always that upon such dispute or difference being so referred the Arbitrator shall be appointed and may, and, if either party so requests, shall give such orders or directions as may be appropriate with a view to enabling the reference to be heard and determined as soon as reasonably practical after Practical Completion of the Works or alleged Practical Completion of the Works or termination or alleged termination of the Contractor's employment under this Contract or abandonment of the Works or to assist the parties in settling the dispute or difference.

.5 No decisions given by the Adjudicator shall disqualify him from being called as a witness and giving evidence before an Arbitrator appointed in accordance with Article 5 and Clause 39.

Dispute or difference—Adjudicated Provisions

S1.4 If any dispute or difference shall arise between the parties in respect of an Adjudicated Provision such dispute or difference shall be dealt with under this Supplementary Provision and Article 5 Clause 39 in the same way as a dispute or difference under any other provision of this Contract but subject to the terms of S1.3.3 in regard to any conflict between an Adjudicated Provision and any other provision of this Contract.

Provisions relating to appointment of Adjudicator

S1.5 If:

.1 an Adjudicator is deceased or is not named in Appendix 1 the adjudicator shall be a person appointed by the appointee named in the Appendix pursuant to Clause 39.1; or

.2 the Adjudicator named in Appendix 1 is unable or unwilling to act as the Adjudicator, the named Adjudicator may appoint an appropriately qualified person as the Adjudicator for all the purposes of S1 or if the named Adjudicator does not so appoint the Adjudicator shall be a person appointed by the appointee named in the Appendix pursuant to Clause 39.1.

Provided that no person shall be appointed or, if appointed, shall act as the Adjudicator who has any interest in the Contract or in any other contracts in which the Employer or the Contractor is engaged unless the Employer, the Contractor and the Adjudicator so interested otherwise agree within a reasonable time of the Adjudicator's interest becoming apparent.

Adjudication—costs

S1.6 In any reference to an Adjudicator each party shall be responsible for its own costs.

S1.7 Before giving his decision to the parties the Adjudicator may require payment of his fee. The Employer and the Contractor shall bear the cost of the Adjudicator's fee in equal proportions.

S2 Submission of drawings etc. to Employer

Employer's Requirements—procedures on submission

S2.1 The Contractor shall comply with any provisions in the Employer's Requirements on
- .1 the submission to the Employer, prior to their use for the construction of the Works, of all (or such as are specified in the Employer's Requirements) drawings, details, documents or information which are reasonably necessary to explain and/or amplify the Employer's Requirements or the Contractor's Proposals; or to enable the Contractor to execute and complete the Works or to comply with any instruction issued by the Employer, and
- .2 the rights of the Employer in regard to his comments on the drawings, details, documents or other information submitted in accordance with S2.1.1.

Effect of Employer's comments or lack of comments

S2.2 Neither the comments nor the lack of comments by the Employer pursuant to the procedure to be followed by the Employer in giving his comments referred to in S2.1.2 shall relieve the Contractor of any liabilities or obligations under the Contract unless the comments specifically so state.

Applications of clauses S3 and S6

S2.3 Clauses S.3 and S.6 shall apply to the drawings, details, documents or information referred to in S2.1.1.

S3 Site Manager

Employer's Requirements—S3 to replace Clause 10

S3.1 Where the Employer's Requirements so state the provisions of S3 shall apply in place of Clause 10.

Provisions on appointment of Manager by Contractor

S3.2 The Contractor shall, prior to the commencement of the Works on the site, appoint a Manager to whose appointment the Employer shall have consented in writing, to act as the full-time representative of the Contractor on the site in charge of the Works. The Contractor shall not remove or replace the manager so appointed without the written consent of the Employer, which consent shall not be unreasonably withheld or delayed, and any instructions given to the Manager so appointed shall be deemed to have been issued to the Contractor.

Attendance of Manager and others—meetings convened by Employer

S3.3 As and when reasonably requested to do so by the Employer the Manager referred to in S3.2 and such other of the contractor's servants, agents, suppliers or subcontractors as may from time to time be necessary shall attend meetings convened by the Employer in connection with the Works.

Keeping of records by Manager

S3.4 The Manager shall keep complete and accurate records in accordance with any provisions relating thereto in the Employer's Requirements and shall make the same available for inspection by the Employer and/or the Employer's Agent at all reasonable times.

S4   Persons named as subcontractors in Employers Requirements

Employer's Requirements—work to be executed by a named person

S4.1 Where the Employer's Requirements state that work (Named Subcontract Work) is to be executed by a named person who is to be employed by the Contractor as a subcontractor (Named Subcontractor) the following provisions shall apply.

Contractor's obligations to enter into Subcontract with named Subcontractor

S4.2 .1 As soon as reasonably practicable after entering into this Contract the Contractor shall enter into a subcontract with the Named Subcontractor and notify the Employer of the date of such subcontract.

.2 If the Contractor is unable to enter into a sub-contract with the Named Subcontractor he shall immediately inform the Employer of the reason for such inability. Provided that reason is bona fide the Employer shall either:

(a) Remover the reason for the inability so that the Contractor can enter into the subcontract by a Change which amends the relevant item in the Employer's Requirements; or

(b) Omit by a Change the Named Subcontract Work from the Employer's Requirements and issue instructions as to the execution of that work.

The relevant provisions of Clause 12 (valuation of Changes), Clause 25 (extension of time) and Clause 26 (loss of expense) shall apply to any Change issued under S4.2.2.

Provisions on Change instructions referred to in S4.2.2(b)

S4.3 .1 The Change referred to in S4.2.2(b) shall not include any requirement for the work to be executed by a person named in the Change instruction but may require the Contractor to select another person to carry out the work subject to the consent of the Employer to the person so selected which consent shall not be unreasonably withheld or delayed.

.2 The Change referred to in S4.2.2(b) may provide that the work is to be executed by a person to whom Clause 29 refers (the Employer or persons employed or otherwise engaged by the Employer).

Provisions on determination of employment of of a Named Subcontractor

S4.4 .1 If the Contractor wishes to determine the employment of the Named Subcontractor for some default, whether by act or omission, by the Named Subcontractor he shall first obtain the consent of the Employer which consent shall not be unreasonably withheld or delayed.

.2 If the employment of the Named Subcontractor is determined the Contractor shall himself complete any balance of the Named Subcontract Work left uncompleted at the date of determination. Such completion shall be treated as if it were work executed in accordance with a Change except where the determination has resulted from the default, whether by act or omission, of the Contractor, or, where Clause S4.4.1 applies, the consent of the Employer has not been obtained as required by that clause.

.3 The Contractor shall account to the Employer for any amounts which he has by reasonable diligence recovered, or which he could by reasonable diligence have recovered, from the Named Subcontractor in respect of the determination as legally due to the Contractor, and which can properly and fairly be regarded as due to the Employer in reduction of the cost to the Employer of the aforesaid Change.

.4 The Contractor shall include in the Named Subcontract conditions a provision which states that the Named Subcontractor, having had notice of the terms of S4 of these Supplementary Provisions, undertakes not to contend, whether in proceedings or otherwise, that the Contractor has suffered or incurred no loss and/or expense or that his liability to the Contractor should be in any way reduced or extinguished by reason of S4 and in particular S4.4.2.

Named Subcontractor obligation of Contractor under Clause 2.1

S4.5 The contractor shall remain wholly responsible for carrying out and completing the Works in all respects in accordance with Clause 2.1 notwithstanding that the Employer's Requirements state that work (Named Subcontract Work) is to be executed by a named person to which the provisions of S4 apply.

S5 Bills of Quantities

Bills of Quantities in Employer's Requirements

If the Works are described in the Employer's Requirements by Bills of Quantities (Bills) prepared by or under the direction of the Employer the following provisions shall apply.

Method of measurement used

S5.1 The Employer's Requirements shall state the method of measurement in accordance with which the Bills have been prepared.

Errors in Bills

S5.2 Any errors in description or quantity in the Bills shall not vitiate this Contract but the error shall be corrected by the Employer and such correction shall be treated as if it were a Change in the Employer's Requirements.

Use of rates and prices in Bills

S5.3   In any valuation under Clause 12.5 a reference to the rates and prices in the Bills shall be substituted for the references to the values of work set out in the Contract Sum Analysis.

Formula adjustment—use of rates and prices in Bills in place of items

S5.4   Where Clause 38 (use of price adjustment formulae) applies the allocation of items and values in accordance with the Formula Rules shall so far as relevant and applicable be effected by reference to the items and to the rates and prices in the Bills in substitution for the reference to the items and values in the Contract Sum Analysis. In Clause 38.2 after the words "Contract Sum Analysis" the following shall be inserted: ", and the Employer shall provide amplification of any Bills of Quantities included in the Employer's Requirements,".

S6   Valuation of change instructions—direct loss and/or expense—submission of estimates by the Contractor

Modification of Clauses 12, 25 and 26

S6.1   Clause 12 (*Changes in the Employer's Requirements and provisional sums*) Clause 25 (*extension of time*) and Clause 26 (*loss and expense caused by matters affecting regular progress of the Works*) shall have effect as modified by the provisions of S6.2 to S6.6.

Instructions under Clause 12—provision of estimates by Contractor

S6.2   Where compliance with instructions of the Employer under Clause 12 will in the opinion of the Contractor or of the Employer entail a valuation under Clause 12.4 and/or the making of an extension of time in respect of the Relevant Event in Clause 25.4.5.1 and/or the ascertainment of direct loss and/or expense under Clause 26 the Contractor, before such compliance, shall submit to the Employer within 14 days of the date of the relevant instruction (or within such other period as may be agreed or, failing agreement, within such other period as may be reasonable in all the circumstances) estimates, or such of those as are relevant, as referred to in S6.3.1 to S6.3.5 unless:
   .1   the Employer with the instructions or within 14 days thereafter states in writing that such estimates are not required; or
   .2   the Contractor within 10 days of receipt of the instructions raises for himself or on behalf of any subcontractor reasonable objection to the provision of all or any of such estimates.

Content of Contractor's estimates

S6.3   The estimates required under S6.2 shall be in substitution for any valuation under Clause 12.4 and/or any ascertainment under Clause 26 and shall compromise:
   .1   the value of the adjustment to the Contract Sum, supported by all necessary calculations by reference to the values in the Contract Sum Analysis;
   .2   the additional resources (if any) required to comply with the instructions;
   .3   a method statement for compliance with the instructions;
   .4   the length of any extension of time required and the resultant change in the Completion Date;
   .5   the amount of any direct loss and/or expense, not included in any other estimate, which results from the regular progress of the Works or any part

thereof being materially affected by compliance with the instructions under Clause 12.

Agreement of estimates

S6.4 Upon submission of the estimates required under S6.2 the Employer and Contractor shall take all reasonable steps to agree those estimates and upon such agreement those estimates shall be binding on the Employer and Contractor.

Failure to agree contractor's estimates—consequences

S6.5 If within 10 days of receipt of the Contractor's estimates the Contractor and Employer cannot agree on all or any of the matters therein the Employer:
.1 may instruct compliance with the instruction and that S6 shall not apply in respect of that instruction; or
.2 may withdraw the instruction; or
.3 may refer the matters not agreed for decision by the Adjudicator under S1
Where the Employer withdraws the instructions under S6.5.2 such withdrawal shall be at no cost to the Employer except that where the preparation of the estimates involved the Contractor in any additional design work solely and necessarily carried out for the purpose of preparing his estimates such design work shall be treated as if it were in compliance with a Change instruction.

Non-compliance by Contractor with S6.2—consequences

S6.6 If the Contractor is in breach of S6.2 compliance with the instruction shall be dealt with in accordance with Clauses 12, 25 and 26 but any resultant addition to the Contract Sum in respect of such compliance shall not be included in Interim Payments but shall be included in the adjustment of the Contract Sum under Clause 30.5. Provided that such addition shall not include any amount in respect of loss of interest or any financing charges in respect of the cost to the Contractor of compliance with the instruction which have been suffered or incurred by him prior to the date of issue of the Final Statement and Final Account or the Employer's Final Statement and the Employer's Final Account.

S7   Direct loss and/or expense—submission of estimates by Contractor

Modification of Clause 26

S7.1 Clause 25 shall have effect as modified by the provisions of S7.2 to S7.6.

Entitlement to amount of direct loss and/or expense—submission of estimate by Contractor

S7.2 Where the Contractor pursuant to Clause 26.1 is entitled to an amount in respect of direct loss and/or expense to be added to the Contract Sum, he shall (except in respect of direct loss and/or expense dealt with or being dealt with under S6) on presentation of the next Application for Payment submit to the Employer an estimate of the addition to the Contract Sum which the Contractor requires in respect of such loss and/or expense which he has incurred in the period immediately preceding that for which the Application for Payment has been made.

Submission of further estimate by Contractor

S7.3 Following the submission of an estimate under S7.2 the Contractor shall for so long as he has incurred direct loss and/or expense to which Clause 26.1 refers, on presentation of each Application for Interim Payment submit to the Employer an estimate of the addition to the Contract Sum which the Contractor requires in respect of such loss and/or expense which has been incurred by him in the period immediately preceding that for which each Application for Payment is made.

Employer's decision on estimate

S7.4 Within 21 days of receipt of any estimate submitted under S7.2 or S7.3 the Employer may request such information and details as he may reasonably require in support of the Contractor's estimate but within the aforesaid 21 days the Employer shall give to the Contractor written notice either
  .1 that he accepts the estimate; or
  .2 that he wishes to negotiate on the amount of the additions to the Contract Sum and in default of agreement *either* to refer the issue for decision by the Adjudicator under S1 *or* to decide that the provision of Clause 26 shall apply in respect of the loss and/or expense to which the estimate relates; or
  .3 that the provision of Clause 26 shall apply in respect of the loss and/or expense to which the estimate relates.

Addition to the Contract Sum

S7.5 Upon acceptance or agreement or an Adjudicated decision under S7.4.1 or S7.4.2 as to the amount of the addition to the Contract Sum such amount shall be added to the Contract Sum and no further additions to the Contract Sum shall be made in respect of the direct loss and/or expense incurred by the Contractor during the period and in respect of the matter set out in Clause 26.2 to which that amounted related.

Non-compliance by contractor with S7.2 and S7.3—consequences

S7.6 If the Contractor is in breach of S7.2 and S7.3 direct loss and/or expense incurred by the Contractor shall be dealt with in accordance with Clause 26 but any resultant addition to the Contract Sum shall not be included in Interim Payments but shall be included in the adjustment of the Contract Sum under Clause 30.5. Provided that such addition shall not include any amount in respect of loss of interest or financing charges in respect of such direct loss and/or expense which have been suffered or incurred by the Contractor prior to the date of issue of the Final Statement and Final Account or of the Employer's Final Statement and Employer's Final Account as the case may be.

TYPICAL CONTRACT SUM ANALYSIS

Typical Contract Sum Analysis
To be completed and returned with tender

ELEMENTS and Sub-Elements

		£	£
1 SUBSTRUCTURE			
(a)	Foundations, including Ground Floor Slab	☐	☐
2 SUPERSTRUCTURE			
(a)	Frame	☐	
(b)	Building Envelope	☐	
(c)	Upper Floors	☐	
(d)	Staircases	☐	
(e)	Internal Walls	☐	
(f)	Permanent Partitioning	☐	
(g)	Semi Permanent Partitioning	☐	
(h)	Demountable Partitioning	☐	
(i)	Moveable Partitioning	☐	
(j)	Toilet Cubicles	☐	☐
3 ROOF			
(a)	Construction	☐	
(b)	Finish	☐	☐
4 WINDOWS AND EXTERNAL DOORS			
(a)	Windows	☐	
(b)	Doors	☐	
(c)	Screens	☐	☐
5 INTERNAL DOORS, SCREENS AND BALUSTRADES			
(a)	Doors	☐	
(b)	Sliding Folding Doors/ Screens	☐	
(c)	Balustrades	☐	☐
	Carried Forward		☐

Typical Contract Sum Analysis (Cont'd)

	£	£
Brought Forward		☐

6 FINISHES

6.1 Wall Finishes
(a) Wall Plaster
(b) Paint
(c) Ceramic Tiling
(d) Wallpaper
(e) Other

6.2 Floor Finishes
(a) Floor Screeds
(b) Sheet Finishes
(c) Tiled Finishes
(d) Carpets
(e) Stairs
(f) Skirtings
(g) Buffers & Handrails
(h) Matwells/ Matts
(i) Other

6.3 Ceiling Finishes
(a) Ceiling Plaster
(b) Plasterboard
(c) Suspended Ceilings
(d) Other

7 FIXTURES AND FITTINGS

(a) Reception Counter/ Screens
(b) Bar Counter and back shelves
(c) Coffee Bar counter
(d) Pinboards
(e) Display Screens

	£	£
Carried Forward		☐

Typical Contract Sum Analysis (Cont'd)

	£	£
Brought Forward		☐

8 SERVICE INSTALLATIONS

8.1 Disposal Systems
(a) Rainwater Systems
(b) Internal Drainage
(c) Sanitary Fittings
(d) Refuse Chutes
(e) Builder's Work in connection

8.2 Electrical Installations
(a) Mains Distribution
(b) Sub-Mains Distribution
(c) Lighting
(d) Power
(e) Emergency Lighting
(f) Heating and Ventilating Electrics
(g) Fire Alarm System
(h) Telephones
(I) Security
(j) Audio Visual Equipment
(k) Lightning Protection
(l) Builder's Work in connection

8.3 Mechanical Installations
(a) Heating Systems
(b) Domestic Services
(c) Thermal Insulation
(d) Air Conditioning
(e) Builder's Work in connection

8.4 Conveyor Installations
(a) Passenger Lifts
(b) Goods Hoists
(c) Builder's Work in connection

Carried Forward

Typical Contract Sum Analysis (Cont'd)

	£	£
Brought Forward		

9 EXTERNAL WORKS

  (a) Hard Surfacing
  (b) Soft Surfacing
  (c) Drainage
  (d) Walls & Planters
  (e) Railings & Gates
  (f) Signage
  (g) External Services
  (h) Service Connections
  (I) Sub-Stations

10 CONTINGENCIES

11 PRELIMINARIES

  (a) Initial Costs
  (b) Time Related Costs
  (c) Completion Costs

12 DESIGN COSTS

  (a) Statutory Charges
  (b) Pre-Tender Design Costs
  (c) Design Development

    TOTAL OF WORKS   To Form of Tender   £

N.B The above is not a standard elemental listing, but was project specific -
being created to cover the known elements arising from the
Employer's Requirements document.

EMPLOYER'S TENDER APPRAISAL

EMPLOYER'S TENDER APPRAISAL

Where: Column A are the Employer's Cost Plan Elemental values, and Columns B and C are the Elemental values from the Contract Sum Analyses submitted by the two lowest tenderers. Gross Floor Area (G.F.A.) : 5,500 Sq.M.

Ref.	Element	Total Cost of Element			Cost/Sq.M. G.F.A.			Percentage of Total Cost		
		A £	B £	C £	A £	B £	C £	A £	B £	C £
1.0	Substructure	275,000	189,655	208,559	50.00	34.48	37.92	0.06	0.05	0.05
1.1	Earthworks	-	8,960	25,027	-	1.63	4.55	-	0.00	0.01
1.2	Piling & Pile Caps	-	120,680	102,194	-	21.94	18.58	-	0.03	0.03
1.3	Ground Slab & Beams	-	60,015	81,338	-	10.91	14.79	-	0.02	0.02
2.0	Superstructure	1,606,260	1,382,885	1,206,075	292.05	251.43	219.29	0.36	0.36	0.31
2.1	Frame	131,888	90,442	52,673	23.98	16.44	9.58	0.03	0.02	0.01
2.2	Upper Floors	281,840	224,445	216,455	51.24	40.81	39.36	0.06	0.06	0.06
2.3	Roof	274,640	201,722	161,413	49.93	36.68	29.35	0.06	0.05	0.04
2.4	Stairs	53,806	45,250	24,801	9.78	8.23	4.51	0.01	0.01	0.01
2.5	External Walls	479,830	331,204	311,821	87.24	60.22	56.69	0.11	0.09	0.08
2.6	Windows & External Doors	222,688	174,230	161,430	40.49	31.68	29.35	0.05	0.05	0.04
2.7	Internal Walls & Partitions	72,352	122,591	98,446	13.15	22.29	17.90	0.02	0.03	0.03
2.8	Internal Doors	89,216	193,001	179,036	16.22	35.09	32.55	0.02	0.05	0.05
3.0	Internal Finishings	381,536	293,093	284,735	69.37	53.29	51.77	0.08	0.08	0.07
3.1	Wall Finishings	159,888	110,964	114,932	29.07	20.18	20.90	0.04	0.03	0.03
3.2	Floor Finishings	140,000	118,392	108,994	25.45	21.53	19.82	0.03	0.03	0.03
3.3	Ceiling Finishings	81,648	63,737	60,809	14.85	11.59	11.06	0.02	0.02	0.02
4.0	Fittings & Fixtures	239,412	221,500	226,763	43.53	40.27	41.23	0.05	0.06	0.06
4.1	General Joinery	34,600	21,000	24,003	6.29	3.82	4.36	0.01	0.01	0.01
4.2	Fixed Furniture	12,312	8,000	10,260	2.24	1.45	1.87	0.00	0.00	0.00
4.3	Provisional Sums	192,500	192,500	192,500	35.00	35.00	35.00	0.04	0.05	0.05
	Carried Forward	2,502,208	2,087,133	1,926,132	454.95	379.48	350.21	0.56	0.55	0.50

EMPLOYER'S TENDER APPRAISAL

Ref.	Element	Total Cost of Element			Cost/Sq.M. G.F.A.			Percentage of Total Cost		
		A £	B £	C £	A £	B £	C £	A £	B £	C £
	Brought Forward	2,502,208	2,087,133	1,926,132	454.95	379.48	350.21	0.56	0.55	0.50
5.0	**Mechanical & Electrical**	887,392	808,623	928,080	161.34	147.02	168.74	0.20	0.21	0.24
5.1	Sanitary Installations	32,640	1,048	49,088	5.93	0.19	8.93	0.01	0.00	0.01
5.2	Disposal Installations	21,760	30,285	19,283	3.96	5.51	3.51	0.00	0.01	0.00
5.3	Space Heating & Air Handling	271,104	266,050	312,558	49.29	48.37	56.83	0.06	0.07	0.08
5.4	Electrical Installations	419,216	390,955	399,171	76.22	71.08	72.58	0.09	0.10	0.10
5.5	Lift Installations	74,672	53,405	69,480	13.58	10.62	12.63	0.02	0.02	0.02
5.6	Builders Work in connection	10,000	3,880	20,500	1.82	0.71	3.73	0.00	0.00	0.01
5.7	Provisional Sums	58,000	53,000	58,000	10.55	10.55	10.55	0.01	0.00	0.01
6.0	**External Works**	270,400	197,892	195,841	49.16	35.98	35.61	0.06	0.05	0.05
6.1	Site Works	110,300	90,941	91,526	20.05	16.53	16.64	0.02	0.02	0.02
6.2	Drainage	63,852	43,409	41,815	11.61	7.89	7.60	0.01	0.01	0.01
6.3	External Services	77,248	44,542	43,500	14.05	8.10	7.91	0.02	0.01	0.01
6.4	Provisional Sums	19,000	19,000	19,000	3.45	3.45	3.45	0.00	0.00	0.00
7.0	**Contingencies**	165,000	165,000	165,000	30.00	30.00	30.00	0.04	0.04	0.04
		3,825,000	3,258,648	3,215,053	695.45	592.48	584.56	0.85	0.85	0.83
8.0	**Preliminaries**	425,000	376,452	468,322	77.27	68.45	85.15	0.09	0.10	0.12
		4,250,000	3,635,100	3,683,375	772.73	660.93	669.70	0.94	0.95	0.95
9.0	**Design Fees**	250,000	190,000	195,500	45.45	34.55	35.55	0.06	0.05	0.05
10.0	**TOTAL CONSTRUCTION COST**	4,500,000	3,825,100	3,878,875	818.18	695.47	705.25	1.00	1.00	1.00

High/Low values indicating potential misunderstanding of the Employer's Requirements

THE RECOMMENDED BCIS ELEMENTS FOR DESIGN AND BUILD

The Recommended B.C.I.S. Elements for Design and Build
To be completed and returned with tender

ELEMENTS and Sub-Elements	£	£
1 SUBSTRUCTURE		▨
2 SUPERSTRUCTURE		
A Frame		▨
B Upper Floors		
C1 Roof structure	☐	
C2 Roof coverings	☐	
C3 Roof drainage	☐	
C4 Roof lights	☐	
C Roof		▨
D1 Stair structure	☐	
D2 Stair finishes	☐	
D3 Stair balustrades and handrails	☐	
D Stairs		▨
E External walls		
F1 Windows	☐	
F2 External doors	☐	
F Windows and external doors		▨
G Internal walls and partitions		▨
H Internal doors		▨
3 FINISHES		
A Wall finishes		▨
B Floor finishes		
C1 Finishes to ceilings	☐	
C2 Suspended ceilings	☐	
C Ceiling finishes		▨
4 FITTINGS AND FURNISHINGS		
A1 Fittings,fixtures and furniture	☐	
A2 Soft furnishings	☐	
A3 Works of art	☐	
A4 Equipment	☐	
A Fittings and furnishings		▨
Carried Forward		☐

This Contract Sum Analysis is reproduced with the kind
permission of the R.I.C.S. Cost Information Service.

The Recommended B.C.I.S. Elements for Design and Build (Cont'd)

ELEMENTS and Sub-Elements	£	£
	Brought Forward	

5 SERVICES

A	Sanitary appliances		
B	Services equipment		
C1	Internal drainage		
C2	Refuse disposal		
C	Disposal installations		
D1	Water - mains supply		
D2	Cold water services		
D3	Hot water services		
D4	Steam and condensate		
D	**Water installations**		
E	**Heat source**		
F1	Water and/or steam (heating only)		
F2	Ducted Warm air (heating only)		
F3	Electricity (heating only)		
F4	Local heating (heating only)		
F5	Other heating systems (heating only)		
F6	Heating with ventilation (air treated locally)		
F7	Heating with ventilation (air treated centrally)		
F8	Heating with cooling (air treated locally)		
F9	Heating with cooling (air treated centrally)		
F	**Space heating and air treatment**		
G	**Ventilating systems**		
H1	Electrical source and mains		
H2	Electric power supplies		
H3	Electric lighting		
H4	Electric lighting fittings		
H	**Electrical installations**		
I	**Gas installations**		
J1	Lifts and hoists		
J2	Escalators		
J3	Conveyors		
J	**Lift and conveyor installations**		

| | Carried Forward | |

The Recommended B.C.I.S. Elements for Design and Build (Cont'd)

ELEMENTS and Sub-Elements	£	£
	Brought Forward	☐
5 SERVICES (Cont'd)		
K1 Sprinkler installation	☐	
K2 Fire-fighting installation	☐	
K3 Lightening protection	☐	
K Protective installations		▨
L Communication installations		▨
M Special installations		▨
N Builder's work in connection with services		▨
O Builder's profit and attendance on services		▨
6 EXTERNAL WORKS		
A1 Site preparation	☐	
A2 Surface treatments	☐	
A3 Site enclosure and division	☐	
A4 Fittings and furniture	☐	
A Site works		▨
B Drainage		▨
C1 Water mains	☐	
C2 Fire mains	☐	
C3 Heating mains	☐	
C4 Gas mains	☐	
C5 Electric mains	☐	
C6 Site lighting	☐	
C7 Other mains and services	☐	
C8 Builder's work in connection with external services	☐	
C External Services		▨
D1 Ancillary buildings	☐	
D2 Alterations to existing buildings	☐	
D Minor building works		▨
7 PRELIMINARIES		▨
8 EMPLOYER'S CONTINGENCIES AND PROVISIONAL SUMS		
A Work complete before commencement of construction	☐	
B During construction	☐	▨
9 DESIGN FEES		▨
	TOTAL	£ ▨

CHANGE ORDER PROFORMA

Change Order Proforma

Project :

Form of Contract : J.C.T. With Contractor's Design 1981
                                  - Supplementary Provisions incorporated.

Contractor :

---

1   **Proposed Change :** ....   **Ref :** ....

     Raised by : ....   **Date :** ....

     Form of Instruction :   ....   **S6.2 Date estimate required**

                                                                    by : ....

     S6.2.1 :   **Estimate not required by Employer (If applicable) :   Yes/No**

                      Date of Employer's written notification :   ....

     S6.2.2 :   **Date of Contractor's written objection (if applicable) :**   ....

---

2   **Contractor's Proposal :**   **Date :** ....

     S6.3.1:   **Proposed value : £**   ....   **Extra/Saving**

                      Reasonable back-up details provided :   Yes/No

     S6.3.2 :   **Additional resources notified by Contractor (if applicable) :**

                      ....

     S6.3.3 :   **Method Statement provided :   Yes/No**

     S6.3.4 :   **Additional Time notified (if applicable) :**   .... **Days**

                      Reasonable analysis provided :   Yes/No

     S6.3.5 :   **Direct Loss & Expense notified (if applicable) :**

                      ....

Change Order Proforma (Cont'd)

Project :

Form of Contract : J.C.T. With Contractor's Design 1981
                   - Supplementary Provisions incorporated.

Contractor :

---

1   Proposed Change :   ....   Ref : ,............

---

3   Employer's Decision :

   S6.5:   Date for Decision :   ....

   S6.3.1 :   Value agreed : £   ....

   S6.3.2 :   Additional resources agreed :   Yes/No

   S6.3.3 :   Method Statement agreed :   Yes/No

   S6.3.4 :   Additional Time agreed :   Yes/No

              Revised Completion Date :   ....

---

4   Failure to agree Contractor's Proposals

   Employer's Options : (One only to apply)

   S6.5.1 :   Proceed subject to Clauses 12, 25 and 26 :   Yes/No

   S6.5.2 :   Proposed Change withdrawn :   Yes/No

   S6.5.3 :   To be referred to Adjudication :   Yes/No

Signed on behalf of the Employer   Date :   ....

....   Issued :   ....

ROLLING FINANCIAL REPORT AND FINAL ACCOUNT

Rolling Financial Report and Final Account

Month : ....

Project : ....

Form of Contract : J.C.T. With Contractor's Design 1981
                    - Supplementary Provisions incorporated.

Contractor ....

A	COST	£	£

Agreed Contract Sum

Less : Provisional Sums
       Contingencies

Add : Agreed S6 Changes
      reported upto end of
      last month

Add : S6 Changes agreed
      this month :
      Ref.   Item

Add : Provision still required for :

      Provisional Sums
      Contingencies

      ANTICIPATED FINAL ACCOUNT

      Representing a potential underspend/overspend
      against the original Contract Sum of :

Rolling Financial Report and Final Account   **Month :** ....

Project :   ....

Form of Contract :   J.C.T. With Contractor's Design 1981
                       - Supplementary Provisions incorporated.

Contractor   ....

B   **TIME**

Agreed Contract Completion Date :   ....

Additional Time agreed and reported
to the end of last month (S6) :   .... **Days (Calendar}**

Additional Time agreed this month (S6) :
         Ref.   **Item**
                                            .... **Days**
                                            .... **Days**
                                            .... **Days**

                              Total   .... **Days**

Revised Contract Completion Date :   ....

C   **COMMENTS**

Financial matters pending :

Anticipated actual Completion Date :   ....

Signed on behalf of the Employer :

....   **Date :** ....

EMPLOYER'S ANALYSIS OF CONTRACTOR'S FINAL ACCOUNT

EMPLOYER'S ANALYSIS OF CONTRACTOR'S FINAL ACCOUNT

Where: Column A are the values from the Contract Sum Analysis, and Column B are the values from the Contractor's Final Account. Gross Floor Area (G.F.A.) : 5,500 Sq.M.

Ref.	Element	Total Cost		Cost/Sq.M.		Change	Comments
		A C.S.A. £	B F/A £	A C.S.A. £	B F/A £	%	
1.0	**Substructure**	189,655	205,330	34.48	37.33	8.27	
1.1	Earthworks	8,960	22,398	1.63	4.07	149.98	Additional Fill - Not authorised
1.2	Piling & Pile Caps	120,680	120,680	21.94	21.94	0.00	
1.3	Ground Slab & Beams	60,015	62,252	10.91	11.32	3.73	Agreed Change
2.0	**Superstructure**	1,382,885	1,715,206	251.43	311.86	24.03	
2.1	Frame	90,442	95,250	16.44	17.32	5.32	
2.2	Upper Floors	224,445	221,780	40.81	40.32	(1.19)	Agreed Change
2.3	Roof	201,722	201,722	36.68	36.68	0.00	Agreed Change
2.4	Stairs	45,250	73,540	8.23	13.37	62.52	Additional Moulds - Rejected
2.5	External Walls	331,204	564,866	60.22	102.70	70.55	Additional Stonework - Dispute
2.6	Windows & External Doors	174,230	175,444	31.68	31.90	0.70	Saving Required - Not Extra
2.7	Internal Walls & Partitions	122,591	189,750	22.29	34.50	54.78	Omits Value understated
2.8	Internal Doors	193,001	192,854	35.09	35.06	(0.08)	Agreed Change
3.0	**Internal Finishings**	293,093	284,735	53.29	51.77	(2.85)	
3.1	Wall Finishings	110,964	210,768	20.18	38.32	89.94	Omits Value understated
3.2	Floor Finishings	118,392	110,980	21.53	20.18	(6.26)	Agreed Change
3.3	Ceiling Finishings	63,737	62,810	11.59	11.42	(1.45)	Agreed Change
4.0	**Fittings & Fixtures**	221,500	226,763	40.27	41.23	2.38	
4.1	General Joinery	21,000	18,650	3.82	3.39	(11.19)	Agreed Change
4.2	Fixed Furniture	8,000	8,650	1.45	1.57	8.13	Agreed Change
4.3	Provisional Sums	192,500	185,660	35.00	33.76	(3.55)	Agreed Change
	Carried Forward	2,087,133	2,432,034	379.48	442.19	16.53	

EMPLOYER'S ANALYSIS OF CONTRACTOR'S FINAL ACCOUNT

Ref.	Element	Total Cost		Cost/Sq.M.		Change	Comments
		A C.S.A. £	B F/A £	A C.S.A. £	B F/A £	%	
	Brought Forward	2,087,133	2,432,034	379.48	442.19	16.53	
5.0	**Mechanical & Electrical**	808,623	1,075,744	147.02	195.59	33.03	General Remeasurement without
5.1	Sanitary Installations	1,048	1,620	0.19	0.29	54.58	reference to authorised Changes
5.2	Disposal Installations	30,285	42,976	5.51	7.81	41.91	- Contractor's F/A submission
5.3	Space Heating & Air Handling	266,050	372,560	48.37	67.74	40.03	rejected by Employer
5.4	Electrical Installations	390,955	481,533	71.08	87.55	23.17	Changes agreed, but overvalued
5.5	Lift Installations	58,405	63,750	10.62	11.59	9.15	Remeasured contrary to C.S.A.
5.6	Builders Work in connection	3,880	20,705	0.71	3.76	433.63	Changes agreed, but overvalued
5.7	Provisional Sums	58,000	92,600	10.55	16.84	59.66	
6.0	**External Works**	197,892	202,057	35.98	36.74	2.10	
6.1	Site Works	90,941	92,550	16.53	16.83	1.78	Saving Required - Not Extra
6.2	Drainage	43,409	43,409	7.89	7.89	0.00	
6.3	External Services	44,542	44,542	8.10	8.10	0.00	
6.4	Provisional Sums	19,000	21,546	3.45	3.92	13.40	Agreed Change
7.0	**Contingencies**	165,000	-	30.00	-	(100.00)	Contingencies fully expended
		3,258,648	3,709,835	592.48	674.52	13.85	
8.0	**Preliminaries**	376,452	495,660	68.45	90.12	31.67	Excluding Disruption Claim
		3,635,100	4,205,495	660.93	764.64	15.69	
9.0	**Design Fees**	190,000	315,000	34.55	57.27	65.79	No basis for Adjustment stated
10.0	**TOTAL CONSTRUCTION COST**	3,825,100	4,520,495	695.47	821.91	18.18	Footprint of Building not changed

Unacceptable result - rejected by the Employer

N.B. On the above Contract the Supplementary Provisions had been incorporated, but the Employer's Agent had failed to insist on the pre-costing requirements of Clause S6 - hence the disaster and consequent dispute.

INDEX